Exploring the
EDINBURGH TO GLASGOW
CANALS

The Union Canal, The Forth and Clyde Canal,
Country Parks and Antonine Wall

Hamish Brown

First published in 1997
This revised edition published in 2006 by

Mercat Press
10 Coates Crescent
Edinburgh EH3 7AL
www.mercatpress.com

ISBN 10: 1-84183-096-8
ISBN 13: 978-1-84183-096-4

British Library Cataloguing in Publication Data
A catalogue record for this book is available from the British Library

Set in Gill Sans and Palatino at Mercat Press
Printed in Great Britain by Bell & Bain Ltd., Glasgow

CONTENTS

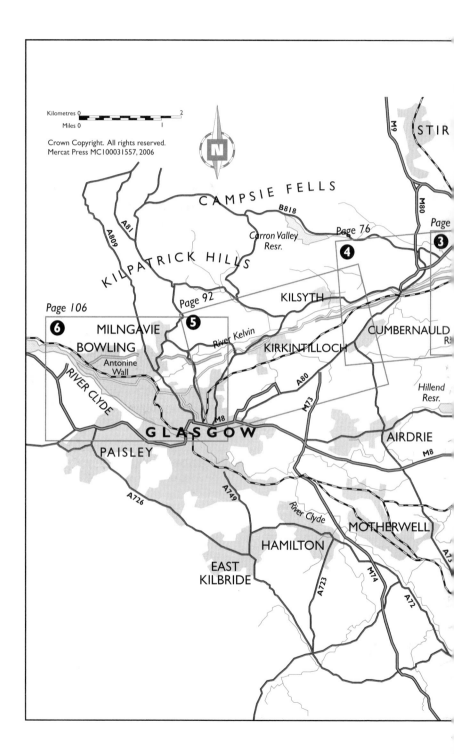

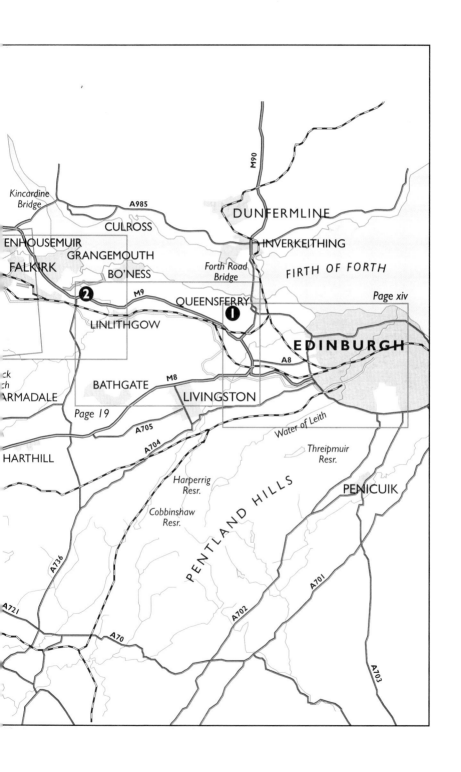

Page xiv

Page 19

INTRODUCTION

The towpaths of the Union and Forth & Clyde canals offer some of the pleasantest and quietest walking in the country and, with historic towns on the route, several country parks and the major fascination of the Antonine Wall, can give the visitor plenty to explore over many visits, or be an excellent holiday trek.

This book describes the canals from east to west in logical progression, so it also acts as a long-distance footpath guide. Situated in the centre of Scotland, the canals can give equally pleasant walking on a sunny summer evening or a winter day when hills and Highlands might be inadvisable. Good stretches are accessible to wheelchair users. Where difficult, car access is described.

Metric or Imperial?

Official notices on the canals can be in metric or imperial, so be prepared for some inconsistencies. Heights from the map are perforce given in metres but most other signs are given as miles—our botched attempt at going metric still not remedied. I'll try and give both.

Maps

Each chapter heading is followed by the numbers of the Ordnance Survey (OS) maps covering that section. OSLR means Ordnance Survey Landranger, scale 1:50,000, and OSE means Ordnance Survey Explorer 1:25,000. The Landranger covers the canals on sheets 66, 65, 64. The more detailed Explorer numbers are 350, 349, 348, 342. However, walking the canals with just one map is possible by using the excellent special GEOprojects/British Waterways map guide *Forth and Clyde and Union Canals (with the Crinan Canal)*. This gives everything one could need. Order in any bookshop or from GEOprojects (UK) Ltd, 9-10 Southern Court, South Street, Reading, RG1 4QS. Tel: 0118-939-3567, Fax: 0118-959-8283.

Opposite: The Seagull Trust's St John Crusader II *returning to Ratho from a cruise.*

The maps in this guide are simple outlines and one of the above should be used in the field.

Wildlife

Despite being a man-made structure, the canals offer a substantial wildlife population of great interest: trees, marsh and water plants and aquatic life as well as the more visual birdlife. (On a busy Glasgow street ducks may be paddling overhead!)

Other Users

Dogs should be kept under control in order not to cause disturbance—or endanger other users, especially cyclists. The canal towpath is a great place to exercise dogs and there are generally no restrictions. Don't allow dogs to foul the towpath. A plea to cyclists, with whom the towpaths are rightly popular—always give warning of your approach, and please be prepared to give way to walkers. Too few of you do either at present! Buy and use a bell!

Practicalities

Over much of the canals there is ready access, often with wheelchair access too. The linear nature of the canals means they can be explored in sections by walkers being dropped and picked up at agreed spots or using the good bus or train options along the route. Best of all, take a week and walk end-to-end with the freedom and interests that gives. Staying locally gives a much greater 'feel' of the country passed—and makes use of the side walks to the attractive country parks that lie on the route.

While this guide is mainly written for walkers, those taking vessels on the canals will still find plenty of useful information and the background stories, I hope, will appeal to all. To avoid repetitions, many features, such as mileposts, stagings, pylons and signposts, are only mentioned where of practical importance or interest.

One sorry observation is best made here, rather than as a constant moan through every urban area, and that concerns the vandalism, litter and graffiti encountered, something on which the authorities or our social conscience fails dismally.

Background History

Much about canals in general, and specific features in particular, appears in the main text, but a few comments may be of interest here. In their heyday, the lack of locks on the Union Canal allowed travellers a speedy crossing between the cities of Edinburgh and Glasgow. The once-thriving service took as little as 13 hours, and cost the equivalent of 7½ pence. At one time there were plans to run the canal through Princes Street Gardens in Edinburgh and down to Leith Docks, but fierce opposition blocked this extension.

The concept and planning of the canals was marked throughout by dissension. The Forth & Clyde Canal opened in 1790, the Union Canal in 1822, but with an inter-city railway opening as early as the 1840s, they then began a century of decline. Only just in time was their amenity and recreational value realised, with various enthusiastic bodies, local authorities and British Waterways Scotland finally receiving the fantastic Millennium funding that has both canals open and working again today. What an impossible dream that would have appeared in 1970.

The Forth & Canal Canal started at Grangemouth in 1768 (it was only fully opened in 1790, as funds ran out during construction) and this is often claimed to be the first canal cutting in Scotland, a fact which, as a Fifer, I cannot let pass unchallenged. The first canal known in Scotland was at Upper Largo, when the great naval hero Sir Andrew Wood (c.1455-1539) had some of his captive English crews cut a canal from Largo House to the local kirk so he could be rowed to church in his admiral's barge. This was at the end of the 15th century.

The first proposal for a canal to link the Forth and the Clyde came from Charles II. He was very interested in naval and engineering progress but, alas, was usually in debt as well. It's an interesting speculation that had he not poured money into the mole at Tangier, there might just have been monies for a canal across Scotland.

The 35 mile (56km) Forth & Clyde Canal was built to take sea-going craft, avoiding the dangers of sailing north about Scotland, and to facilitate east-west and Atlantic trade. The width of the canal was set at 28ft (8m) and the depth at 8ft (2.7m).

Work on the Forth & Clyde Canal began in 1768 by the engineer John Smeaton. Five years later it had been completed from the Forth to Kirkintilloch, which immediately became a 'port'! Another two years saw it reach Maryhill, but then the money ran out. In 1777 Glasgow merchants got the branch as far as Hamiltonhill. In 1785, with money raised on the forfeited Jacobite estates, Robert Whitworth, an engineer, built the then extraordinary Kelvin Aqueduct and took the line to Bowling on the Clyde, a five year effort. The Glasgow branch was pushed through to Port Dundas.

The 31½ mile (51km) Union Canal was nicknamed The Mathematical River, partly because it followed the 240ft (73km) contour, and also because it maintained a regular width (35ft/11m) and depth (5ft/1.5m). While on figures, in 1834 no fewer than 121,407 passengers travelled the canal. Meals, music, even gaming tables were provided to pass the time, and there was a night sleeper service which was popular with both business-men and honeymoon couples. Darwin as a student wrote to his sister about going to Glasgow by canal in 1826.

Linlithgow Palace was garrisoned as the local people were doubtful about the wild Irish navvies on their doorstep. Two have gained deserved notoriety through a second career as body-snatchers: Burke and Hare. Ironically, much of the expertise gained on building the canals was then used in the construction of the railways that led to the death of the canals. The enormous work done in restoring the canals has provided many ideas and techniques for canal restorations elsewhere. The Falkirk Wheel was one unique concept and went instantly from inspiration to icon.

Horse-drawn barge

It is hard to envisage the Scotland of pre-canal times, when the only transport was four-footed, either as riding and pack animals or pulling inefficient carts over an almost impossible landscape. With Glasgow and Edinburgh growing and both domestic and industrial demand for coal reaching a critical stage, there had to be some new development. The Forth & Clyde Canal was a splendid pioneering venture and, not for the first time, Edinburgh spent years squabbling over routes and plans before agreeing on the Union Canal. (I notice that the Edinburgh provost, who bitterly opposed the finally accepted plan of Hugh Baird, has his name down among the subscribers.) John Rennie's plans were thus superseded. Thomas Telford had supported Baird, who consulted him on various occasions.

It is also hard to envisage the sheer scale of the work. The whole length was divided into lots and these were allocated to various contractors. It was all pick, shovel and wheelbarrow work, employing thousands of itinerant labourers. The squalor can be imagined. A satellite picture of the time would have shown a dirty brown scar across Scotland. The scar would hardly have healed before the railways were making the canals redundant.

Look out for details of construction. Bridges on the Union Canal are numbered from Edinburgh and there are 62 of them, plus several new ones and, of course, all those where the canal does the bridging, which is often much more exciting. Tow-ropes have often worn grooves into stonework, there are mile posts, wharves, winding-holes for turning boats, intakes and spillways and other features on the bridges.

Restoration

In 1994 British Waterways announced they would seek funding from the new National Lottery opportunity. That the project finally flew the flag of 'The Millenium Link' hides the years between as hopes rose and fell and many people worked their hearts out on behalf of the dream. Donald Dewar, Secretary of State for Scotland, announced the funding guarantees in 1998, all £78.4 million. This was the minimum needed and millions are still being raised, and spent, as the work of improv-ing and regeneration continues. There were plenty of places with eyesore industrial failures which are now landscaped or developed tastefully—and that effort will continue. How quickly the work settles in never fails to astonish. Do read Guthrie Hutton's book mentioned below for the last chapter that tells the saga of restoration and wonder at the pictures of work in progress: a mammoth undertaking. It all looks so natural now, but we must not forget. The first edition of this book ended with a promise of great things to come, now we see the fulfilment of those hopes. Scotland has a girdle of gold across her waist.

Contacts

I am more of an enthusiast than an expert on the canals and am often delighted, giving slide lectures, to have new information coming from listeners, many who are keen members of local bodies and have supported the canals from the gloom to glory days. A list of clubs and societies is given at the end and new members are always welcome. I am also greatly in debt to the serious historians who have done the demanding research and given us the benefit in words and pictures. One title I would recommend especially is Guthrie Hutton's *Scotland's Millennium Canals* (Stenlake Publishing 2002). A fuller bibliography is also given at the end of the book. Various pubs, restaurants, boat charters and other commercial bodies are also listed at the back. It is always advisable to phone in advance to check on opening hours, etc.

The Leamington Lift Bridge

Mackay Seagull, *one of The Seagull Trust's cruise boats at Ratho*

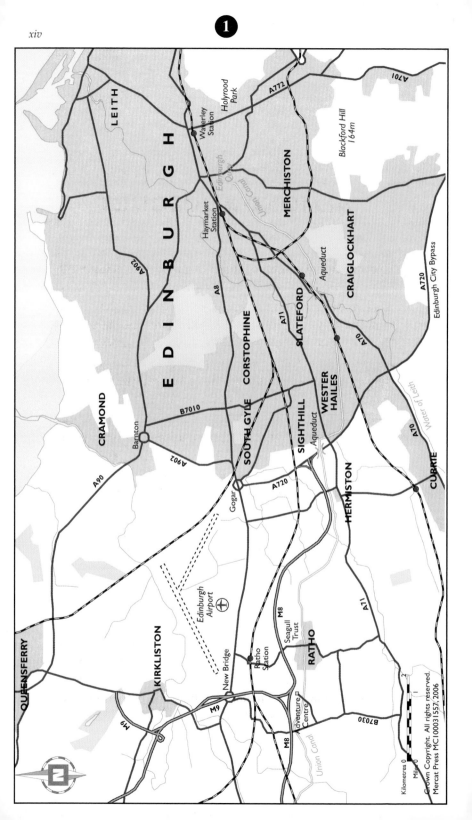

1
EDINBURGH TO THE ALMOND AQUEDUCT
OSLR 66, 65; OSE 350

Lothian House on Lothian Road, a five minutes walk from the west end of Princes Street, has a mural depicting a horse-drawn canal barge on the Union Canal and the wording 'Here Stood Port Hopetoun'. To me this has always been the poignant symbol of times past, but walk on another five minutes and the present canal terminus, Edinburgh Quay, is reached. This is just one of the latest developments following the rebirth of the canals across Scotland, a landmark for times to come.

Lothian House was built in 1922 on the site of the basin named after the Earl of Hopetoun, a major early investor whose collieries supplied much of Edinburgh's coal. Off at a tangent near the end (reaching to Morrison Street) lay the coaling basin of Port Hamilton, another name that nods to a grandee investor. Another basin, Lochrin Basin, lay south of Fountainbridge and served the brewery—which is still there. The basin had closed, even before Port Hopetoun and Port Hamilton were filled in and 'developed' in the 1920s. By then the whole area had become seedy and neglected. A new basin, just the stump of the amputated original, was then called Lochrin Basin.

With the Millennium Link project restoring navigation coast to coast and city to city (the Forth & Clyde Canal completed in 2001 and the Falkirk Wheel/Union Canal in 2002) ambitious plans were made for revitalising both the Edinburgh and Glasgow city centre termini of the canals, work which will continue for some years yet.

Edinburgh Quay, Terminus of the Union Canal

Walking down Lothian Road keep on the east side to best see the mural on Lothian House, very much a Twenties building with a touch of art deco, then head along Fountainbridge. There are odd survivals from the old days: Fat Sam's is based in the 1884 Edinburgh Meat Market, there's Port Hamilton Tavern and the 1859 St Cuthbert's Cooperative Association building with its various mottos, a somewhat strange Victorian survivor among the continuing wild geometrics of modern buildings. Just before the roundabout the building with the Cargo Restaurant gives access onto Edinburgh Quay. If coming from Haymarket Station, head up Morrison Street then right along Gardner's Crescent to reach this roundabout and Edinburgh Quay. Edinburgh Quay is well hidden among the big modern edifices which give this area of the city a unique architectural flavour.

The start of the towpath runs alongside McEwan's Fountain Brewery to the eye-catching Leamington Lift Bridge. This originally stood near Port Hamilton but was moved here and is still in good working order.

This rump of water was once the scene of a headline-grabbing accident. George Meikle Kemp, the creator of the Scott Monument, was a shy, rather odd character whose design was only chosen because of an impasse in awarding the contract to the first choices. He was the compromise candidate. More joiner than architect, his design incorporated ideas ranging from Melrose Abbey to Rouen Cathedral, but then, Scott himself was constantly stealing ideas if not actual structures as he built Abbotsford, his house near Melrose.

Kemp had been to see his contractor (the stone presumably coming on the Union Canal) and left to walk home along the towpath on a cold, dark, foggy night. Somehow he lost his bearings and walked off the pier near the Lochrin Distillery. A week later the first evidence of a tragedy appeared when his stick and hat floated to the surface. He is buried in St Cuthbert's, the church down in the dell below the meeting of Princes Street and Lothian Road. Others buried there include artist Alexander Nasmyth (whose bridge you will see on the River Almond) and the drug-addict writer Thomas de Quincy.

The towpath is on the north bank of the wandering Union Canal throughout and the bridges are numbered on the keystones: 1 to 62, although some bridges have been replaced and new ones have been given numbers like 6A, 6B and so

At Bridge 1 on the Union Canal

on. The first bridge we come to, Viewforth Bridge, has a carving of a symbolic castle over the keystone on the Edinburgh side and Glasgow's fish and tree over the western keystone. There are several utilitarian iron bridges (without numbers) before Bridge 4 (Meggetland Bridge) is reached and the numbering takes on its real interest. Sadly, the last few years has seen a marked increase in graffiti on bridges and walls.

This first mile always has swans, ducks, coots and moorhens who seem oblivious to human activity, unless rushing for bread thrown by residents of the modern flats overlooking the canal on the south bank. A bridge is passed (no access to/from towpath) and then older tenements overlook the canal.

The canal curves and the green sweep of Harrison Park appears on the right. The next colourfully painted iron bridge (Harrison Road) has the massive red sandstone Polwarth Parish Church beside it and beyond lies the Forth Canoe Club boathouse and a variety of hiring and cruising craft including the cruise and drive restaurant longboat *Zazou*. The Ogilvie Terrace Moorings has a launching slip, winding hole and all the facilities for cruising: water, electric hook-up, rubbish disposal etc. The big iron bridge, with spiky finials, carries Ashley Terrace and is inscribed 'Lockhart Bridge 1904'. Continuing, the houses stand back from the banks and the feeling is more rural.

A rather weathered milestone is the next landmark, just before the Edinburgh University boathouses on the far bank, and then the canal bestrides a railway line which is backed by a view of Corstorphine Hill with its masts (and Edinburgh Zoo). There is also a canal overflow on this bank.

There's a staging before the Meggetland pair of bridges, the first a modern concrete horizontal structure, the second is an example of the original standard Union Canal bridges style: inscribed 'Bridge 4'. Boroughmuir RFC lies off right. Housing lines the canal's south bank, but the north is being developed. There's another almost illegible milestone then a footbridge/pipeline combination linking Allan Park Road on the north and Craiglockhart on the south.

Slateford Aqueduct in early days (© Edinburgh City Library)

The canal wiggles along its contour to reach the next landmark, the Prince Charlie Aqueduct, which was rebuilt in 1937, one of the better concrete bridges of that vintage when seen from Slateford Road, which it spans. The name perhaps comes from the Pretender setting up camp close by in 1745 to await the surrender of the city. A flight of steps leads down for shops or bus. Continuing, industrial works lie below to the right while the Pentlands loom on the left front, but then comes the most spectacular structure within the city, the eight-arched Slateford Aqueduct, 152m (500ft) long and 23m (75ft) above the Water of Leith, with the canal carried in an iron trough. This fine aqueduct, Scotland's second largest, is now rather hemmed in by the busy Slateford Road to the south and a lower, but fine, arched railway bridge to the north. There is an overspill at the east end of the aqueduct, the overflow spraying down into the Water of Leith. Hugh Baird is the engineer responsible for our trio of great aqueducts, this one marginally bigger than the Almond Aqueduct.

Just before the aqueduct a flight of steps descends to reach a sturdy footwalk passing under the aqueduct and viaduct—part of the Water of Leith Walkway. The nearby

Water of Leith Centre, besides information and interpretive displays, has a café, the first so accessible since leaving Edinburgh Quay. The River Leith rises in the Pentlands and can be followed afoot (or cycled) right through the city to the sea, offering further circular walks from the towpath.

A stump of milestone marks the west end of the aqueduct and the towpath runs parallel with the A70 Lanark road to an iron girder of a bridge with the far abutment strapped up. There's a signpost and a sculpted sign for the National Cycle Network (75) and a note of the Millennium Acres Community Woodland, then the first of the tree sculptures: of two men, one with a spade, the other holding a rock, carved from a tree in situ. Railings mark a low, arched pedestrian underpass up to the A70 (Kingsknowe), then the railway crosses the canal. Bridge 5 is in standard MM style, then playing fields are passed to reach Wester Hailes, marked by huge tower blocks on the south side of the canal. Hailes Park was once a quarry 100ft (30m) deep. Much of the stone went to London.

Wester Hailes—modern geometrics on the Union Canal

The canal was culverted for about a mile through this big housing development, so cutting it again was another major undertaking, carried out with the usual flair of the Millennium Link restorations. The new crossings that were needed have played havoc with the numbering of bridges. The next, after an overspill, is 5A, and is twinned with an older bridge which does not have a number. To work past the dominating, quite stylish tower blocks, we have Bridges 6, 6A, a black horizontal footbridge, and 6B (all in MM form). Beside another tree sculpture seat of a frog, there's a long section of moorings. Here the towpath is tarmac and bridge piles on bridge again: a high arched footbridge, another horizontal black one, and Bridge 6C, beside which is a dock with a dragon sculpture.

The ground falls away on the right for a stretch, more moorings (unused, alas, with no security from vandalism) lead to Bridge 7 then, with playing fields passed, there's a last black footbridge and an overspill. And the frenzied roar of Edinburgh's ring road traffic. The canal re-emerged here during the years of culverting and now swings right—the Murray Burn goes under—to pass a row of bollards (Calder Quay) and the last tree sculptures, of otters this time. A concrete flyover (the A71 Livingston/Kilmarnock road) replaces its close neighbour, an original Bridge 8, then there's a MM Bridge 8A (all these liberally paint-daubed) and the canal swings left through the Hermiston Gait development to reach the Scott Russell Aqueduct which crosses 170ft (56m) over the ring road, a noisy fascination which will continue awhile as the 1995 M8/M9 link sweeps parallel to the winding canal, audible always even if not visible. A milestone at the aqueduct neatly indicates 5 miles back the way—and 26½ to go. The aqueduct was constructed in 1987.

Scott Russell (d.1882), was an engineer and naval architect who discovered what was called the Solitary Wave Theory on the canal here. When his craft stopped he noticed a wave went on ahead, unchanged, a long way. This fascinated him and led to the wave-line principle of ship construction and, in the 1960s, to application with fibre optics.

Bridges 9 and 10 soon follow, now cut off from traffic, the latter with a mooring on the south side (where I once came on a dozy cormorant), then Gogar Station Road Bridge 10A, built pre-millennium (1995) with stone abutments and a shapely concrete arch with white railings. Bridges 11 and 12 soon follow, the latter taking a small road north over the M8. Nearing Bridge 13, Jaw Bridge, the Gogar Burn flows underneath and, all along this stretch, kicking stones and old milestones appear. Kicking stones are stones set on the edge of the towpath and designed to give the horses a safer purchase when straining to pull barges along. There's a spillway as we near Bridge 14, Gogar Moor Bridge. This has a small road and, thereafter, a golf course runs along the south side for most of the way into Ratho.

At the start of the village there's a graceful iron seat with a panel showing canal wildlife, then a small perching seat which commemorates the nearby stage marker with a notice '1832. To Edinburgh 7 miles, fare 6d. [2.5p], To Falkirk 25 miles, fare 1/6 [7.5p]'. Another old milestone stands beside the stage post for stages one and two. Across the canal and on to Bridge 15 in Ratho is a long reach of moorings, well-used. There are one or two more attractive resting seats (which continue beyond Bridge 15) and then a picnic area with a superb model of a horse drawn barge in the old days.

In its heyday, Ratho had 14 pubs. The Pop Inn, next to the canal, is reputed to have had a door at each end of the building so the bargeman in charge of the towing horse could enter by one door, enjoy a pint *en passant*, and exit by the other door

The Bridge Inn and the boats of the Edinburgh Canal Centre, Ratho

without having fallen behind his charge. In about 1845 Ratho House was turned into a distillery, the annual production of 42,000 gallons being largely consumed locally. Canal work was obviously a drouthy business. Ratho was actually a change house and fare stage for tolls.

One of Edinburgh's regular exports via the canal was manure (horses rather than horse-power in those days) and this led to great fertility on the canalside farms. Maybe this lingers on, as two local farms hold world records for wheat production. Coal (and building stone) was the big import to the city, and the major reason for the canal being built. The passenger side was never lucrative and was soon killed by the railways.

The Bridge Inn was originally a farm, then a staging post on the canal, now it is renowned for the excellence of its food, the friendly service and delightful setting. It is also the base of the Edinburgh Canal Centre, which offers a variety of cruises with the two restaurant longboats, the *Pride of the Union* and *Pride of Belhaven*. These often go along to the Almond Aqueduct where people dine on board ('in the air' it feels) or al fresco on the bank. There are even cruises to Santa's Secret Island in the festive season, which were once seen on Blue Peter.

The Seagull Trust is also based here, and runs trips for disabled passengers in the *Mackay Seagull* and *St John Crusader II*. Their first cruise took place in 1979; now 5,000 people are involved annually and there are other boats based at Falkirk, Kirkintilloch, and Inverness.

Ratho is a neat little village and, before continuing, do have a look at the old cemetery round the church. Cross the bridge again (Baird Road commemorates the canal's engineer) and the church lies behind the war memorial. Left of the door is an unusual single stone shaped like a coffin; on its side the inscription indicates the incumbent suffered 'an instantaneous death' from a stroke by a thrashing (*sic*) machine (early 19th century).

Ratho is a major access point and has car parks opposite the church and by the Bridge Inn. From the Newbridge

The coffin-shaped gravestone in Ratho churchyard

roundabout turn off for Newbridge, then, almost at once, turn off left and follow Edinburgh Canal Centre signs. Leaving Edinburgh by the A71, the turning off by Dalmahoy Road is similarly signposted.

Back on the canal, you pass the old change house and set off along a good towpath, away from the importuning ducks. Walk along the Ratho Hall grounds with some of Ratho's newer houses over to the south, then there is a peaceful, deep-set, woody section with mature trees shading both banks. As the canal bends right a bridge is glimpsed ahead and at that point, on the right, rough steps will be seen—with a glass roof peeping over the trees—which leads up to the Ratho Adventure Centre. Ironically there's still a sign saying 'Quarry, Keep Out'. This huge (five-stories-high) quarry has been roofed over to create the largest indoor rock-climbing facility in the world, besides much else. A spectacle not to miss. Hopefully the linking steps will be improved. The bridge seen ahead is the new road in to the Adventure Centre. There is a café and restaurant.

Finding the place by road is not easy. Though known as Ratho the access is nowhere near Ratho and the best signposting is from the hectic Newbridge roundabout, with brown 'touristy'-style signs. You turn off for Newbridge, then immediately left. Keep to this road which winds up below successive railway bridges, the M8, and the canal. The signposted drive is clear and crosses the canal again to the Adventure Centre. From here a huge waterfilled quarry can be seen south of the canal which runs in its quiet woody cutting. From the A71 out of Edinburgh turn onto the B7030 at Wilkieston, continue through Bonnington and on to the entrance (right). From Ratho take the road for Bonnington and turn right onto the B7030.

Once out of the trees there is quite a contrast, for the M8 swings alongside the canal for a while to share its roar and bustle with you. There is a wider section, Wilkie's Basin, with an island in it (Santa's Island in season for Ratho cruises), then you cross the B7030 road (Bonnington Aqueduct) from Newbridge to Wilkieston, on an aqueduct which was reconstructed in 1978. Timbers dredged up in the basin are thought to be from one of the 'Swifts' which carried passengers from Edinburgh to Falkirk in three and a half hours. Pulled by two horses (changed periodically) they had prior right of way and charged along in style. The timber baulks stored by the aqueduct allow the canal to be dammed off for repair or maintenance work. The M8 has small tunnels under it for the use of badgers.

After a spell without bridges, numbers 16, 17 and 18 come in quick succession. By the attractive setting of Bridge 16, Nellfield Bridge, you may spot the narrow boat *Thomas Telford*. There is also a milestone and you are back in mature woods again, a gentle spot. Bridge 17 is a high arch with Clifton Hall School to the north, Bridge 18 has a rustic feel (with railings), then, quite dramatically, you reach the high Almond Aqueduct. There is little warning: you round a bend and are on it. On the other side there is a basin, wharf and car park, and the major feeder for the canal. Upstream lies Almondell and Calderwood Country Park, which is well worth a diversionary visit and will be described in the next section.

The car park opposite can be reached from Newbridge, taking the B7030 which passes under the railway, M8 and the Bonnington Aqueduct. Follow 'Union Canal' signs thereafter. There is access to the aqueduct from the car park so even the disabled can see the spectacular setting. Long flights of steps however link the two sides of the canal. On the wall by the towpath is the date 1821. The drive passing under the canal is private, going to Lin's Mill. There is no mill now, but Lin's Grave is hidden away in the woods of the grounds and is inscribed 'Here lyeth the dust of William Lin richt heritor of Lins Miln who died in the year of the Lord 1645'. He was one of the many victims of the plague which ravished Scotland that year.

The boats cruising from Ratho often turn at this spot after sailing out and back over the aqueduct. They may also enjoy a buffet supper ashore and I saw one wedding charter thoroughly enjoying life. An 1834 handbill, offering ten-mile (16km) trips for sixpence (2p) noted that here, "fruits, confectioneries and varieties of refreshment can be had". Now, sadly, litter louts just leave their chip supper debris behind.

2
ALMONDELL AND CALDERWOOD COUNTRY PARK

OSLR 65; OSE 350

This attractive park has the main Union Canal feeder running down through it, so is of particular interest for those walking the canal. The feeder is visible on the south side of the canal beside the staging and winding hole, immediately before the Almond Aqueduct, but to reach it means negotiating a set of steps down (follow the sign for: Country Park) and up again under the aqueduct. Ratho restaurant boats and Seagull Trust sailings often tie up at this corner, after briefly going onto the aqueduct, so passengers can also inspect it on foot.

Walk out from the canalside car park, and after about 100m there are steps (signposted, right, for Almondell Country Park) which drop down to the feeder aqueduct, here going into a final tunnel to reach the canal basin. Turn left and walk up the glen. Across the way Illieston Castle stands boldly against the skyline, and downstream, as one progresses, lies the canal aqueduct, a splendid, buttressed structure. Illieston Castle is a well-preserved tower house, built by John Ellis in 1665 but with earlier buildings going back much further. James II and James IV both used it as a hunting lodge.

The path along by the feeder is clear ,so needs little detailing. There are plenty of gates and stiles and interest in seeing how streams are led under the feeder. Neat bridges bring farm tracks down to riverside fields and periodically the feeder vanishes into conduits, tunnels cut because of spurs in the wending glen. At one spot the river runs hard against the flank so the path heads up, only to descend again by steps. Shortly after, the houses of Shiel Mill lie below and the tarred road to them is crossed by stiles.

Where the feeder comes out of another tunnel a steep path goes up, left, signed Larchwood Walk. Drop down to the graceful single tower suspension bridge across the River Almond. It was built by the Royal Engineers in 1970 and received a Civic Trust Award. In 1986 it was named the Mandela Bridge.

A grassy area, once the old walled garden of Almondell House, leads to the park's Visitor Centre buildings, housed in the stable block of the former mansion which had a connection with the 18th century Erskine family. The 11th Earl of Buchan had lived at nearby Kirkhill House and a brother Henry had built the Almondell mansion. The earl died with no issue to inherit, so much of the Kirkhill contents came to Almondell and Henry's son inherited the title in 1829. Another brother, Thomas Erskine, was a famed forensic lawyer. Henry Erskine himself was a famous lawyer—he twice became Lord Advocate. He started building Almondell in 1790 to his own eccentric design. It was a disastrous enterprise, but he loved the setting which we enjoy today, although we have the additional benefit of the trees in their splendid maturity. The house was damaged by fire in the 1950s and demolished by the army in 1969. Its site, which you pass later, is marked by a parking place for disabled drivers.

Mathematician, astronomer, antiquarian and scholar, the Earl of Buchan constructed a scale model of the Solar System in and around the grounds of Kirkhill House in 1776. The model consisted of the Sun, Mercury, Venus, the Earth and its moon, Mars, Jupiter and its four moons and Saturn with its rings and five moons. Mars is now known to have two moons, Jupiter eleven moons and Saturn nine, and since the construction of the model, Neptune, Uranus, Pluto and more have been discovered; but even bearing this in mind, the model was extraordinarily accurate. It was constructed to a scale of 12,238.28 miles to an inch (approximately 1:775 million), the Sun being represented by a stone sphere six feet in diameter, and the Earth by a bronze sphere 0.646 inches in diameter placed 645 feet (196m) away. The larger planets were made, like the Sun, of stone, while the smaller planets were of bronze.

The model has disappeared, but a summary of the calculations which enabled the Earl to construct the model are preserved on a stone pillar which Buchan erected in the grounds of Kirkhill

The astronomical pillar in the Almondell Country Park

in 1777. The astronomical data inscribed on the pillar still largely holds true today. Buchan also included on the pillar a prediction of the position of the planets on 20 May 2255 (why he chose this date is unknown). The pillar was surmounted by a bell tower, on top of which was a metal cross. After Buchan's death, the bell tower was removed and taken to his younger brother's estate at Almondell, where it was placed in front of the stables over a well (which is still there). The pillar remained at Kirkhill, but by the late 1970s it had collapsed. When the shell of Kirkhill House was sold for private restoration, the stones of the pillar were taken into safekeeping by West Lothian History and Amenity Society, and it was decided to rebuild the pillar, with bell tower and cross, in front of the Visitor Centre at Almondell. And there it is today.

The Visitor Centre (opened in 1981) is a lively, friendly place with interesting historical and wildlife displays, an attractive aquarium and an audio-visual presentation. It is the base for the park ranger service. Soft drinks, tea, snacks and chocolate are on sale if you need refreshments—with benches outside and a snug conservatory within.

Walk on up the drive through the disabled car park. There are some fine specimen trees, including chestnut, copper beech, cypress, red cedar and sequoia, but most eye-catching are the lime trees with great bunches of suckering growths round their bases—good shelter for wintering birds! Most garden birds will be seen, and also wren, treecreeper, woodpecker, robin, wagtail, dipper and mallard, with goosander, cormorant and herons fishing in the river, perhaps on occasion a kingfisher. Bats, roe deer, fox, badger, squirrel, rabbit and even otter have been noted.

A small bridge is passed (Dell Bridge, 1784) and, not long after, the drive swings over a large, ornate, double-arched bridge across the River Almond which was designed by the portrait painter and garden landscaper Alexander Naismith and built about 1800. Nasmythe (there are other spellings as well) is most remembered for his iconic portrait of Rabbie Burns, thought to be the only authentic likeness of the poet. The bridge was recently restored as the various stone tints indicate; it had partly collapsed in 1973.

Don't cross the bridge but continue up the west bank of the Almond. Curiosity will wonder what a pulley system across the river to a small building is for. It is a SEPA (Scottish Environmental Protection Agency) gauging station where regular information is collected and river quality monitored. Once a month samples are taken from the river, usually by just wading out but, in a spate, the samples can be collected by using a gondola along the wire!

Continuing, the path comes to the green arched ironwork of an unusual bridge—it carries the canal feeder stream across the River Almond and has a walkway on top of that so, as I heard an excited child say, "Look! I'm walking on water". It dates to 1820, part of Hugh Baird's canal work.

Still keeping to the east bank, the next structure over the feeder and river is a high (23m/75ft), nine-arched 1885 Camps Viaduct, which once led to mines, brickworks and oil industrial sites at Pumpherston and Uphall. The roofed structure below it, like an extended church porch, is thought to have been built as a precaution against anything dropped off the viaduct.

Instruments in the feeder and a sluice gate at a footbridge are more SEPA recording/control works. By crossing the bridge it is possible to climb steps up onto the viaduct, an airy perch to view the glen. Continuing by the feeder, the East Calder Sewage Works are seen, built in 1960 on the creation of Livingston New Town which lies to the west of this rural world, a sprawl of 40,000 inhabitants now. (Hence the SEPA sites.) Take the small bridge (signposted as a cycleway) across the feeder and gain the sturdy footbridge (Pipe Bridge, 1960) over the Almond. This is the end of our exploration for, upstream, we see the weir across the river which creates the start of the canal feeder. As many as two million gallons a day may be added to the canal. The river starts away to the west near Harthill (the well-known watershed when motoring the M8) and eventually flows into the Forth at Cramond. A reservoir, Cobbinshaw, was built in the Pentlands to provide extra water, which then flows down the Bog Burn, joins the Muirieston Water, then Linhouse Water before joining the Almond. The river flowing under the Almond Aqueduct is 22m (70ft) lower than the canal, hence the length of the feeder which seldom flows more than 2 mph. The feeder channel here is full of monkey flower, bittersweet, greater willowherb (an aquatic cousin of the rose-bay species which is also plentiful) and blue water mint.

The Calderwood part of the park lies further on upstream, but is perhaps best left for a visit when the canal does not claim all our attention, so return to the bridge taking the feeder across the river, 'walk on water' across

and down to the Nasmith Bridge. Cross it and take the
riverside path which can be followed to the Mandela
Bridge. On the way, at the Dell Bridge burn, is a stone
slab inscribed 'Margaret Countess of Buchan dedicated
this forest to her ancestor Simon Fraser 1784', the Simon
Fraser who supported Robert the Bruce. Alternatively,
from the Nasmith Bridge the drive can simply be taken
back to the Visitor Centre. Either way, thereafter it only
remains to wander back down the feeder path to the canal
basin by the great aqueduct, completing a visit of many
interests beside the canal associations.

Swans at Lock 4 on the Forth and Clyde Canal

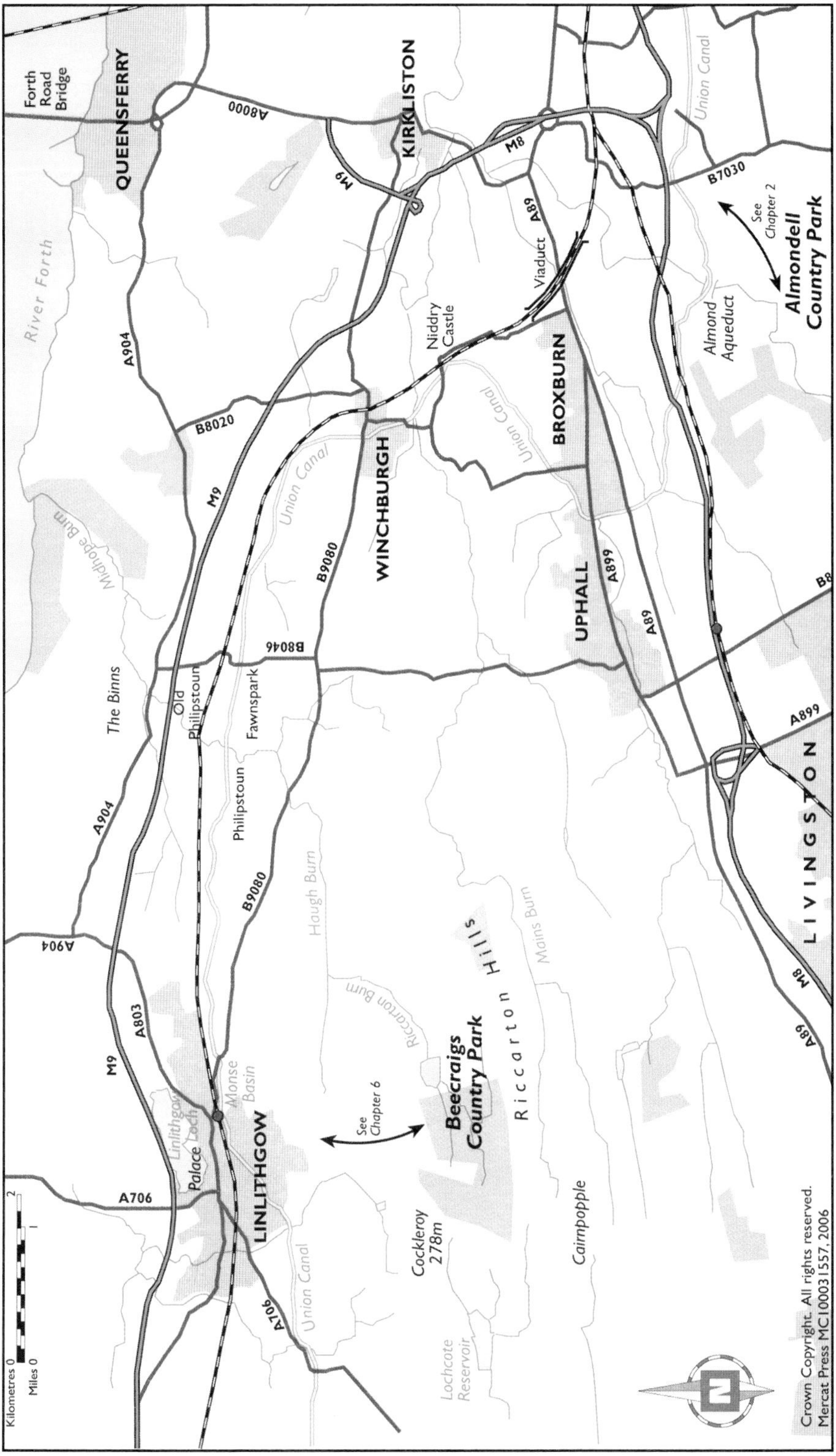

QUEENSFERRY

Forth Road Bridge

River Forth

A8000

KIRKLISTON

M9

M8

A89

Viaduct

Niddry Castle

BROXBURN

Union Canal

B8020

WINCHBURGH

M9

B9080

UPHALL

A899

A89

Union Canal

B7030

See Chapter 2

Almond Aqueduct

Almondell Country Park

B8

A899

LIVINGSTON

M8

A89

Midhope Burn

The Binns

Old Philipstoun

Fawnspark

B8046

Philipstoun

B9080

A904

A904

Hough Burn

Riccarton Burn

Mains Burn

Riccarton Hills

Beecraigs Country Park

See Chapter 6

M9

A803

Linlithgow Palace Loch

Manse Basin

LINLITHGOW

A706

Cockleroy 278m

Cairnpopple

Union Canal

A706

Lochcote Reservoir

Kilometres 0 ___ 2
Miles 0 ___ 1

N

3
ALMOND AQUEDUCT TO WINCHBURGH

OSLR 65; OSE 350

Our third largest aqueduct is impressively high and exposed with the narrow, cobbled pathway (420ft/128m) edged on the left by the iron trough of the canal. On the right, an airy iron railing does little to hide the 70ft (22m) drop to the River Almond. In spate the flat island downstream can be covered with water. The aqueduct is 420ft (128m) in length. There is a superb view north, down the river, to Telford's Almond Valley Viaduct on the main Edinburgh-Glasgow Railway. This was built in 1842 and, with 36 arches, somewhat surpasses the aqueduct's mere five. The viaduct contains over a million cubic feet of masonry. Never intended for today's weights and speeds, it has suffered somewhat and is strapped up to support the big arches. An impressive sight and A-listed.

The control sluice in the middle of the south side of the aqueduct tends to dribble water and, in severe winters, this overflow has been known to freeze solid, creating a remarkable pillar of ice, as for a two month spell in 1895, recorded in the photograph opposite. At the west end of the aqueduct there is a milestone marked 10 (to Edinburgh) and 21 (to Falkirk). About 100m further on an overflow channel, lined with granite setts, crosses the towpath.

Bridge 19 (Broomflats Bridge) is just for a farm track but the minor road bridge, 20, starts a more interesting stretch, with a railway bridge ahead. The railway is the old Edinburgh-

Opposite: the famous 1895 icicle on the Almond Aqueduct (© LUCS)

The Almond viaduct on the Union Canal

Uphall-Bathgate-Coatbridge-Glasgow line, which breaks off from the main Edinburgh-Glasgow line west of Ratho (at Newbridge Junction), and was originally built in 1849 to Bathgate, being extended to Airdrie and Glasgow in 1879. It served the coal, iron and shale oil industries, and the last passenger train was in 1956. The creation of Livingston New Town failed to provide new passengers, and the line west to Airdrie was lifted in 1982. Bathgate station was gutted by an arsonist. Out of this dismal history came resurrection, and the line to Bathgate, an unusual modern station, is now a busy commuter route. Planes from Edinburgh Airport climb steeply into the sky nearby.

Squeezed between railway and M8, Bridge 21 (Kilpunt Bridge) just gives rural access, then we face the frantic motorway. When the canal was built there was no M8, of course, and its development caused a major blockage and set a problem for any reinstating of the canal. The M8 was slightly raised, and by making the S-bend the canal passes under the motorway. (You can see where the original line went straight on.)

Bridge 22 just serves Learielaw Farm, a nice name. (Others hereabouts include Loup-o-Lees, Birdsmill, Pumpherston,

Powflats and Lookaboutye.) The next Bridge, 23, Drum-shoreland Bridge, has a cluster of buildings, staging and working boats and is the Union Canal works base of British Waterways Scotland.

The building against the bridge (on the towpath side) has odd eye-shaped windows, and the corners of the bridge have grooves cut by towropes during the busy past.

Bridge 24 has what looks like Christmas decorations hanging overhead, to stop swans flying into the wires, and then, on the far side, there's a spillway into the Ryal Burn. On the right there's a row of cottages with gardens running down to the burn. The next bridge is a 1930s ugly red brick and iron structure for the modern A89, and Bridge 25 (Miss Margaret's Bridge) links housing schemes with steps onto the towpath as you head north, cross the Brox Burn, and reach the A899, the main Uphall-Broxburn road, at a modern concrete bridge of no character. Turn right if you want Broxburn's town centre and, at the traffic lights in the centre, if not returning to this spot turn left again to rejoin the canal as it leaves the town.

Broxburn has a wide range of shops, coffee houses and pubs, and is a friendly place despite a rather unpretentious

The Almond Aqueduct as it used to be. (© LUCS)

appearance. It straggles along north of the Brox Burn, and grew rapidly with the shale oil industry. In 1861 the population was 660, in 1891 it was 5,898. With that industry gone, there is an air of survival only. They were either affy wild or unco guid in Victorian times. I lost count of the churches along the long High Street that joins it with Uphall. The West Church is 'weird and wonderful Gothic', and the Roman Catholic church the real showpiece. St Nicholas Parish Church, on the B8046 out of Uphall, is the only old church. Uphall was once Strathbrock (valley of badgers) and Broxburn is from the same old word brock for badger (badger stream). Uphall is a name which seems to cause pronunciation problems for some reason. It is just as written, Up-hall, but people will produce Uffle and such like.

The Earl of Buchan built his solar-system model at the family home, Kirkhill. The family were connected with the area until after the Second World War. Broxburn's Roman Catholic church was built in 1880 for the Dowager Countess of Buchan. She presented it with the font, which has had a varied history. Dating to pre-Reformation times, it was ejected from the new kirk and for some time was used as a cattle water trough on a local farm. When it was recognised for what it was, the farmer gave it to the countess.

The A899 bridge leads to the wide Port Buchan basin, with seats, landscaping, picnic tables and toilets etc. Sheltered housing makes this a pleasant corner, then the canal swings to the northeast, between housing and various works, to reach Greendykes Road Bridge (27) which had to be rebuilt —in the Millennium style—to open the canal again. The red shale bings are now immediately ahead.

On setting off, a big launching slip and car park is passed. The next two bridges are basically derelict, but the bing across the water bears the marks of adventurous bikers although surprisingly there are likely to be tits working through the alder and willow planting, moorhens fussing in the canal, swans gliding by and foxgloves colonising the slopes beyond. A high percentage of the foxgloves are white, maybe due to the poverty of the soil or its chemical composition. Further along the canal are banks of scented stock. (In September the slopes of the bing chitter with the sound of explosive broom pods.) A canal is often an artery of life in an otherwise dead landscape. The brick-red colour of the bings is hardly surprising, for much of this spoil has been turned into bricks, or used for land reclamation at Grangemouth, or for motorway construction. The word 'bing' has its derivation in the Gaelic 'ben', meaning a hill.

Shale oil manufacturing was a typical Victorian enterprise. James 'Paraffin' Young first came to West Lothian in search of 'cannel coal' (candle coal—used for lighting as it burnt with such a bright flame), and this led him to develop a process to extract paraffin oil and wax from the oil-bearing shales. So the oil industry began here. At its peak there were 120 works employing 13,000 people, but by 1873 the number had dropped to 30 as the oil wells of the USA began to produce their black gold. Young died in 1883, and the last works closed in 1962. 'Paraffin' Young was a chemical engineer from Glasgow. A fellow student who became a lifelong friend was David Livingstone, and much of the sponsorship for the latter's travels came from Young.

Oil shale bings near Broxburn

Queen Victoria had Falls named after her, but there is a branch of the Lualaba named Young River. In this quiet setting (aircraft permitting), it is hard to imagine the atmosphere 150 years ago when dozens of chimneys poured smoke into the air. Grangemouth is quite modest in comparison to pictures of the old oil industry. The Almond Valley Heritage Centre in Livingston has a museum on the shale industry (as well as a mill, working farm, etc) and is worth a visit.

There's a pleasant open stretch with the bing passed. Away on the left is another which is jokingly referred to as Ayers rock. Niddry Castle lies ahead, below yet another bing mountain. Winchburgh can be reached by walking past the castle as well as by keeping to the towpath, so a brief note on both.

As the canal swings left (with Bridge 30 in sight) leave the towpath for a minor road. This crosses the main Edinburgh-Glasgow railway line and a glimpse over the parapet shows a cutting far deeper than the canal's. The line then goes through a tunnel under Winchburgh. Turn left once over the bridge ('Footpath to Winchburgh' sign) on the drive leading to the Castle. This bold tower sits somewhat incongruously in the middle of a golf course (with some modern defences

Niddry Castle which stands near the Union Canal.

against golf balls rather than cannon balls). The castle has had impressive restoration. It was built by the Seton family in 1490. Lord Seton was one of those who helped Mary Queen of Scots escape from Loch Leven Castle, and she was brought to Niddry, briefly, before the battle of Langside led to her final flight and imprisonment in England. One of her 'Four Marys' (attendants) was Mary Seton. Mary, Queen of Scots, was three times as long a prisoner of Elizabeth as she was a free monarch in Scotland. The castle was sold in 1676 and abandoned early the next century.

Skirt the castle on the right (east) to drop down onto Niddry Castle Golf Course, and cross it to walk up to Winchburgh on a path outside the perimeter of the golf course and below the shale 'mountain'. The path comes out at the car park for the golfers and the road up to the main street passes a couple of classic 1890s miners 'rows', the cottages no doubt with a car at every door, changed days indeed.

En route, you can turn right to climb the bing through Hank Hill Wood, and it is worth wandering up this artificial hill for the view, and to see how nature is slowly greening-over the barren waste. They are beginning to be positive, rather than negative, features in the landscape. I trust some will be allowed to survive, both as wildlife sanctuaries and monuments to an important industry. Winchburgh had a brickworks which used the shale waste and was sending loads into Edinburgh by canal up till 1937—the last real commercial use of the canal.

Keeping to the towpath to reach Winchburgh, the next bridge, 30, requires some caution from motorists, being humped, narrow and with sharp bends. The Niddry Burn comes out from under the canal, having risen near Beecraigs Country Park and entering the Almond at Kirkliston. Bell's Mill Wharf Bridge, 31, is the home of the 'Bridge 19-40 Canal Society' and is dated 1820. A path angles up to the road into town but is best ignored. A deep cutting follows with a modern pedestrian bridge linking housing on the west bank with the town centre. Bridge 32 carries the B9080 road to Linlithgow and has had a tubular footbridge added on the south side. Go through under the bridge and steps angle up to the road. Turn left into the town.

On the left, 120 metres on, is the Tally Ho, a substantial red coloured building. Inside this inn are many photographs of Winchburgh in the days when the oil and brick works were in full operation. On the other side there is a delightful war memorial figure of a drummer boy. An architectural guide points out 'in the middle, a pompous Police Station'—now a pharmacy. A carved bird and the date 1903 adorn the Star and Garter. A few shops offer the chance of supplementary snacks.

Broxburn Shale Oil works, now completely gone. (© Guthrie Hutton)

4
WINCHBURGH TO LINLITHGOW

OSLR 65; OSE 350, 349

As you leave Winchburgh, a signpost indicates that Linlithgow is 6 miles ahead, the Falkirk Wheel 17½ and Edinburgh 15½. Many consider the miles to Linlithgow the most serene and pleasing of all. Much of the way is tree-lined and the noise of parallel rail and M9 lines is somewhat muted. There is a large car park down past a house at the start.

A cottage on the far side has landscaped its portion of canal, Bridge 33 (Myre Bridge) only leads a track over to fields, and in the deep-set stretch beyond there are obvious kicking stones and the stage post for sections two and three, along with a battered old milestone. Bridge 34 just carries a track on a concrete bridge built on original abutments, with tracks everywhere used by local walkers. Over it, off the path up to the Winchburgh-Linlithgow road, lie the ruined gables of ancient Auldcathie Church. The name Priestinch points to a long-vanished pre-Reformation parish hereabouts.

Bridge 35, Craigtoun Bridge, is rather grander with railings and ornamentation as befits a drive from the heyday of the Hopetoun estate. Even the number 35 is pushed to one side of the keystone. Bridge 36 is an estate bridge too, and from it a track goes up a steeply inclined bridge to pass over the railway. I always find it interesting to climb up to every bridge possible. There are kicking stones again by the sturdier Bridge 37, which goes 'nowhere to nowhere'.

Humpback Bridge 38 at Fawnspark has been strengthened and given traffic lights to cope with heavy commuter traffic.

Fawnspark (Bridge 38) from up on an old shale bing

There's a car park beside it, but the exit is onto a blind bend; highly dangerous. A winding hole lies opposite and the farm beyond breeds Clydesdale horses; some may be seen grazing in the fields. The canal then runs deeply through what feels like a cutting, as the waste from the shale oil industry is piled up on both sides, a wooded area much enjoyed by mountain bikers. Abutments show where a bridge connected with the industry once stood. Bridge 39 is now only used by cross-country walkers (there is a path from the Philpstoun road south to Threemiletown), but is worth going up onto it for an attractive view, including a major overspill just west of it. Good open, rural walking continues all the way to Linlithgow.

Easy to overlook is a short aqueduct section of canal, for a road from Philpstoun passes under the canal. A mere hamlet now, Philpstoun was a place entirely dependent on the oil industry. The name dates back to Philip d'Eu, a 12th century Norman who was granted land here. You may notice a scent of aniseed. It comes from the feathery leaves of Sweet Cicely (*Myrrhis odorata*), which grows all along the canal. So do brambles—a bonus for an autumn tramp. Two access paths lead in from the village and, on leaving, bites of semicircular masonry mark a water-filled underpass linking fields.

After a slow transformation, the canal, from being deep in its private jungle, now runs along high, open country with fantastic views over the rolling Lowlands to the swelling Ochil's along the northern skyline. The tower on the hill to the north is above The Binns, home of Tam Dalyell, the long-serving local MP. An ancestor of the same name was a General and a feared persecutor of the Covenanters. He was captured at the battle of Worcester, later reorganised the Russian army for the Tsar, won the battle of Rullion Green (1666) and raised the Scots Greys in 1681. They wore a grey cloth imported from the Netherlands, which the general ordered to try and make his men less conspicuous in the field (a use of camouflage that was not in general practice for another 200 years!) His portrait shows stern features and a huge white beard. He was also notorious for never wearing boots.

Overlooking the Forth Estuary beyond The Binns is ancient Blackness Castle and the palatial Hopetoun House, all worth visits. South of Bridge 40 is Campfleurie House, a French name. Slots under the bridge can have baulks of wood inserted to dam off a length of canal—as does Bridge 42.

Kingscavil's Park Bridge, number 41, was the site of one of the change houses along the canal. The name means the 'Kings plot of land', but the house and estate once belonged to the Hamiltons. Young, newly-wed, cheery Patrick Hamilton was burnt at the stake as an early martyr of the Reformation. Prince Charles slept in the old house in 1745 while his army lay at Threemiletown (Scots miles, longer than English). The bridge's number 41 has almost weathered away, and there are deep rope grooves. There is plenty parking and the attractive Park Bistro is across the road. Fishermen encountered along this stretch, and even fishing within the Edinburgh or Glasgow city boundaries, may land perch, roach, pike, bream, tench, carp or eels, all of which are found in the canals, testimony to their clean water.

As you near Linlithgow, you are passing below Pilgrim Hill, and the name St Magdalene was once that of a fair and hospice on the town's outskirts. The town has spilled out eastwards in a huge, impersonal suburb, Springfield. After Bridge 42, however, you have the Palace and St Michael's Church in view. The canal passes over the B9080 as the town is reached (the bridge has black and white railings). Below is what looks like a distillery with its pagoda-like towers, features which have been carefully preserved in turning the one-time St Magdalene distillery into luxury modern houses. The 1960s saw much of historic interest swept away by an unimaginative local council, so this is a contrast. Liking or loathing is the reaction to the 1964 aluminium crown of thorns on top of St Michael's Kirk beside the Palace.

You come to the bowed parapet of another lane going under the canal. Immediately below is the station, then the Regent complex which stands on the site of the old Nobel Works built in 1701, the Explosives Factory as it became in the world wars. ICI purchased the works and then closed them down in the 1960s.

The grassy area below the castle is called the Peel, there being such a defence long before stone castles were built. Originally the canal had hoped to make use of the loch, but the need to keep to its contour prevented that. At Bridge 43 you reach the large Manse Basin.

Manse Basin, Linlithgow, base for the Linlithgow Union Canal Society.

Here too the corners of the bridge have been deeply worn into grooves by towropes. The canal comes to life with a tearoom, museum and a collection of boats, the creation of LUCS, the Linlithgow Union Canal Society, an enthusiastic body of volunteers who have done much to revitalise this section of the canal, tidying the area, upgrading the towpath, rescuing everything old and interesting, running trips in the *Victoria* (a replica steam packet), the *St Magdalene* (an all-electric vessel) and other old boats, hiring rowing boats and so on. The museum, in the old stables building, is a fascinating record of the canal's past, and there is a short audio-visual presentation. The sprawl of Linlithgow below the canal is worth some exploration, as is the hill country above—these are covered in the next two chapters. The description of the continuation westwards along the Union Canal resumes in chapter 7.

5
LINLITHGOW, TOWN OF BLACK BITCHES

OSLR 65; OSE 349

"When, on a summer evening about the hour of eight, I first beheld Dreamthorp, with its westward-looking windows painted by sunset, its children playing in the single straggling street, the mothers knitting at the open doors, the fathers standing about in long white blouses, chatting or smoking; the great tower of the ruined castle rising high into the rosy air, with a whole troop of swallows skimming about its rents and fissures; when I first beheld all this, I felt instinctively that my knapsack might be taken off my shoulders, that my tired feet might wander no more, that at last, on the planet, I had found a home. From that evening I have dwelt here, and the only journey I am like now to make, is the very inconsiderable one, so far at least as distance is concerned, from the house in which I live to the graveyard beside the ruined castle."

That comes from a book written about 170 years ago by Alexander Smith, who is better known for a second book *A Summer in Skye*. His Dreamthorp name, coined for the town, gave Linlithgow one of its nicknames rather than anonymity. Smith died in 1867 when only 37.

Defoe thought Linlithgow 'a pleasant, handsome, well-built town', where bleaching linen was the obvious industry. Robert Burns was less flattering, saying Linlithgow 'carries the appearance of rude, decayed, idle grandeur' but thought it 'sweetly situated'. William and Dorothy Wordsworth stopped off for breakfast on the way

Linlithgow from the canal

to Edinburgh and the Borders at the end of their Highland Tour but made no particular comment.

For centuries Linlithgow was an important leather-making centre and, like Selkirk, could be somewhat smelly. All the traditional industries have gone, though they are commemorated in the various guilds with their deacons. Linlithgow still elects a Provost, and the people of the town, regardless of sex, are Black Bitches. A black bitch appears on the town's coat of arms. There are actually two coats of arms, the second portraying St Michael having a go at the dragon.

King David I built a house here in the 12th century, but like most Border or Central towns it suffered from the visits of English armies. They burnt the town in 1424. But with the Stuarts came prosperity. James I began the building of the palace and most of his successors added to it. Mary Queen of Scots was born in the palace in 1542. When James VI became James I in 1603 and flitted to London, this became a neglected second home.

Cromwell used it as barracks for nearly a decade. It hosted Bonnie Prince Charlie, but was finally gutted after being occupied by Butcher Cumberland's troops. In 1989 Linlithgow celebrated the 600th anniversary of receiving its

The old doocot near the canal
basin, Linlithgow

charter from Robert II, at the time when, in England, the Black Prince's son was king and Chaucer was penning his tales.

The following basic walkabout takes in most worthwhile features. From Manse Basin you look over Strawberry Bank to a garden with a 16th century beehive-type doocot holding 370 boxes. It has the standard stone courses sticking out to prevent rats climbing up, and a tiny door. Pigeons were popular in medieval times as they provided fresh meat in winter. Only nobles (Ross of Halkeld in this case) were allowed such, the common people just had old salted beef. There were no root crops then for winter feeding, so each autumn animals were slaughtered, or driven south to English markets.

Turn east from the garden to go down a road, barred to traffic, which leads to the High Street after passing under the railway and by the station (on the Edinburgh-Glasgow/Stirling lines). The station was modernised in 1985 and opened by the well-known local MP Tam Dalyell. There's a colourful painting in the upper hall worth seeing. Linlithgow had an ancient right to levy tolls, which it did on roads and then on the canal. The railway, however, refused to pay, and despite various courts upholding the town, the House of Lords finally favoured no tolls for railways.

The road comes out onto the High Street near the one time High Port, beside the Star and Garter Hotel, a black and white confection rather like an English coaching inn. Head left along the High Street, a hotch potch of styles and periods with an unfortunate flat-roofed 'stopper' to the east. A weathered figure of St Michael with the town's coat of arms stands on one of the many old wells, dated 1720 and inscribed 'St Michael is kinde to straingers'.

Across the High Street are several 16th and 17th century houses. The Hamilton Lands were restored by the National Trust for Scotland in 1958. Two have gables facing the street with steep red pantiles. Crowstep gables were designed to allow beams to be placed across a roof which was too steep for ordinary construction. (Behind one of the houses is an old outdoors baker's oven, from the days when fire was a serious hazard.) On the wall at the back of a pend is an inscription 'Ve Big Ye Se Varly 1527'('We build you see warily'). At one time the town house of the Cornwalls of Bonhard stood here. An Alexander Cornwall was said to have been one of six knights dressed as the king—and all, including the real James IV, killed at Flodden.

Across the road, the bulky Victoria Hall, completed in 1889, once had big Gothic towers and pepperpot turrets, but the building has failed steadily, the towers gone, turned into a cinema, then bingo hall, amusement arcade and now boarded up. On the south side still, above the sign of The Four Marys, is a tablet commemorating a Dr Waldie, who introduced chloroform to Sir James Simpson and the medical profession.

Continuing briefly along that side, the council building has a splendid example of a Provost's lamp. The next building, the Sheriff Court, has a tablet commemorating the assassination of the Regent Moray in 1570. Unfortunately it manages to spell Moray incorrectly and also gives the wrong date. This murder was one of the first-ever such deeds using a firearm, and was very carefully prepared. Hamilton of Bothwellhaugh, after firing the shot, made his escape to the continent and, cashing in on the deed, became a professional hit-man. The Archbishop, whose house he had used, was hung.

The town centres round the Cross Well. Cross, weekly market and gibbet have long gone, and the well has had a chequered history, being rebuilt in 1659 after damage from Cromwell's troops. In 1807 it was completely rebuilt, copying the old design, the work being done by a one-handed stonemason. The original Town House was also destroyed by Cromwell, but rebuilt by the king's master mason John Mylne in 1668. Fire damaged it in 1847, when its Italian-style arched portico was replaced by popular wrought ironwork. The present double stairway superseded this in 1907. The well-stocked and helpful Tourist Information Centre is housed here and, west of the Cross, Annet House, 143 High Street, a Georgian merchant house, now houses the Linlithgow Story museum with its life-sized models, memorabilia and video shows, with a restored garden in the rear 'rig'.

The Kirkgate leads up from the Cross to reach the Palace gateway. Panels above have the painted and gilded coats of arms of the Orders of the Garter, the Golden Fleece, St Michael (all conferred on James V) and the Thistle (which James V is thought to have founded). The porch at Abbotsford was based on this entrance—Walter Scott also cribbed bits of Melrose Abbey and Stirling Castle, apart from accumulating original old features.

Through the arch, on the right, is St Michael's Parish Church, a large, cathedral-like building with a long history. Dedicated in 1242, most of it dates to a rebuilding after the 1424 fire. Its main fame is perhaps the window tracery, nota-bly in the St Katharine's Aisle where James IV saw the ghost (a put-up by his worried wife?) who warned of impending doom if he marched an army south to Flodden, as it proved. The Reformation took its toll of the decorative statuary. Cromwell actually quartered his troopers and their horses inside the church, and it accommodated Edinburgh students during the plague winter of 1645-46. There's a mortsafe lid still lying on the south side.

The Creation Window is dedicated to the leader of the 19th century Challenger Expedition, which explored the world's oceans, and the huge window depicts a vast range of animals, birds and fish, including a bright red lobster! Most heart-rending is the story behind the Child Samuel window, which commemorates the little daughter of a previous minister who died when her hair caught fire as she dried it before a blazing fire in the Manse. A sister had previously died when she went through the ice while skating on the loch with her fiancé.

Linlithgow Palace's most famous tale is its capture for Robert the Bruce by a local farmer William Binnie, who regularly used to deliver hay to the garrison. One day he hid men under the hay and stopped the cart in the entrance so the portcullis and gates couldn't operate. More men rushed in, and the palace was won.

The gem of Linlithgow Palace (looked after by Historic Scotland) is its courtyard fountain which has been fully restored: tiers of symbols and figures exquisitely carved—as are the recent replacements. The setting too is grand. It was Mary of Guise who declared she had "never seen a more princely palace".

The Palace stands in what is collectively called The Peel, which, with Falkland Palace, are unique royal parks. To walk round the loch—which gives fine views to palace, church and town—takes about an hour. Two of the loch's islets are artificial, being crannogs (ancient loch dwellings of about 5,000 years ago). Yellow water lilies have flowered in the loch for centuries and it is a bird sanctuary, sometimes with wintering swans gathering in hundreds. At the period when the palace was being used, gardens, orchards and an apiary surrounded it and the loch would provide water for brewing as well as fish, eels, ducks and swans for eating. When Bonnie Prince Charlie marched in the fountain was spouting wine! The town once had about ten wells, hence the jingle:

> Glasgow for bells
> Linlithgow for wells,
> Falkirk for beans and peas
> Peebles for clashes and lees.

The West High Street has less historic interest but provides several more cafés and restaurants with which the town is well supplied. If followed to the end, Preston Road, running south, leads up to the Union Canal at the restored Bridge 45.

6
BEECRAIGS, COCKLEROY
AND CAIRNPAPPLE

OSLR 65; OSE 349

South of Linlithgow lie the Bathgate Hills, as they tend to be called, and they have three notable sites worth visiting, either afoot from Linlithgow or, quite feasibly, by taxi: a country park (with a red deer herd), a notable wee hill with a big view and one of Scotland's finest prehistoric sites. The last really needs one's own transport, but is briefly mentioned here, while Beecraigs Country Park and Cockleroy give some verticality to the contoured level of the canal!

The best approach is up from Bridge 45, Preston Road Bridge, reached from the west end of the High Street or by walking along the towpath from the Manse Basin. When Preston Road reaches the edge of town there's a sign indicating 'Beecraigs 1.2 miles' and a wide route for walkers, cyclists and horse riders has been provided, running parallel with the road to give safe access to the Hillhouse Woods and Beecraigs. There are plenty of gates to allow farm access to fields, and when the route turns left, eastwards, a map board describes the woodland stretching ahead. The wood is a mix of conifer and larch trees with some deciduous along the edges and masses of willowherb.

Follow the walkers' symbol, once over the first rise turning right, past lichened rocks of a quarry, and twisting up a craggier slope to reach an open verdant mossy area. The views back are on a grand scale. The next stile has a 'dog hole' beside it. At the next stile, do not cross as the walkers'

Opposite: the secretive loch in Beecraigs Country Park

symbol suggests, but walk on, past a stand of ash, to merge with a track going through a wall into mature beech trees. The path leads out onto a road at a signed junction for Beecraigs.

Walk down past the restaurant (open for lunches, dinners) and caravan site entrance and the Visitor Centre is off left, a log cabin structure with a grass roof. Minimal refreshments are available. Local displays and park information. From the far corner of the car park there is a raised walkway giving a view over the deer farm fields and off to the green swelling of the Riccarton Hills—not to be taken for Cockleroy. The path goes through the fields, so there's a clear observation of the red deer. They are not tame animals or pets but are farmed commercially just like cattle or sheep, or like the trout at the fish farm beside Beecraigs Loch where you come out. Rainbow trout are reared, may be fed, and purchased for supper. Turn left in the wood to follow the shore round and over the dam, below which lies the fish farm.

The park has a wide range of activities: field archery, orienteering, fishing, cycling, horse-riding, climbing wall, etc. The loch is well hidden among conifer woodlands with Dagger Island covered in Scots pine. Ducks galore, coot, moorhen, greylag, swan and others are listed on a board nearing the west end of the circuit. Exit across a minor road to Lochside car park.

From the top of the car park take the 'Balvormie' footpath up through the trees (the paths can be confusing) to eventually reach the big Balvormie car park (tarred) beside another minor road. Cross it and pass a pond to find the Cockleroy path, which heads off from a wooden toilet block. You reach yet another minor road and the Cockleroy car park. From the far end the Cockleroy path is signposted, to a dark tree tunnel, so there is a sharp contrast when you suddenly come out to the breezy open hillside beyond (where there's a gate/stile). A steep five-minute ascent lands you on the summit of Cockleroy with its view indicator, trig point and 360-degree panorama which rates among the best in the Lowlands. Not bad for 278m (912ft).

View indicator on Cockleroy

The wide saddle of the Cauldstane Slap in the Pentlands, the rock fin of Binny Craig, the crouching lion shape of Arthur's Seat, the hunky stump of the Bass Rock, all lie to the east. Working anti-clockwise, the Forth Bridges are well seen, the Lomonds are bold, behind Linlithgow are the Cleish Hills, then the long horizon of the Ochils. The big, high chimney is that of the Longannet Power Station. Grangemouth is a jungle of chimneys and cooling towers, and beyond is the Kincardine Bridge. (The dusk view when all these areas are lit up is quite spectacular.) Running away to the west are the Campsie Fells, while real Highland hills can also be seen: Ben Vorlich, Ben Ledi, Ben More (by Crianlarich), Ben Venue, Ben Lomond. Balancing the Bass Rock in the Forth, a clear day may reward with a view of the ragged peaks on the Isle of Arran in the Clyde Estuary.

Nearer, below you to the south-west, is the obvious Lochcote Reservoir. Not so obvious, but between it and craggy Bowden Hill to the north, lies a flat area which was a loch until drained last century. A 'crannog' (lake dwelling) was discovered then. Just to be different, Beecraigs Loch is man-made, built by prisoners of war during World War I. Both Cockleroy and Bowden Hill (229m/770ft) summits are the sites of prehistoric forts, though there is nothing much to see. Quite magnificent, however, is the excavated and preserved multi-period prehistoric henge, circles and tomb on Cairnpapple Hill, due south, next to the obvious relay mast—one of the Top Ten prehistoric sites in Scotland. The knobbly nature of these hills points to their volcanic origins.

The grassy depression on Cockleroy is called Wallace's Cradle. The patriot reputedly used the hill as an observation post and safe spot in the dangerous Lowlands, where all his days were spent. He held a parliament at Torphichen, another historic site well worth a visit.

Cockleroy is not a French hybrid word but murdered Gaelic, *cochull-ruadh*, the red cowl. (On an 1898 OS map it is Cocklerue and an 18th century guide had Cuckold le Roy.) This waist-of-Scotland, not surprisingly, is a real mix of Gaelic, Norman and old British names—and some guid Scots ones too, like Burghmuir or Cauldhame. Beecraigs is also Gaelic in origin, from *beithe* (pronounced bey) meaning birch tree.

Retrace the route as far as the Balvormie car park then head left outside the forest to pass (or scramble on) a rope tower before following the track down through the forest back to the Park Centre.

To reach Cairnpapple Hill by car, turn left on leaving the Centre, then take two right turns in quick succession (at North Mains and South Mains) to pull up westwards. First left (a bad bend) and you're soon there. The view is even finer than from Cockleroy, and the henge/cairn, 4000 years old, is fascinating. If the custodian is present you can climb down into the burial chamber of the central cairn.

Afoot, and returning to Linlithgow from Beecraigs, the route can be slightly varied. Return to the Hillhouse Woods and follow the walkers' signed path along through the attractive beech avenue. Level with a gate onto the road you turn right to descend on a grassy path, forking left at a junction. A bench is passed and the path comes out to a tarred road, where great care should be taken as cars often

tear round the hairpin bend 50 metres down. As the road
swings right to make the hairpin, a footpath continues
straight down, with a wall to the left. A water reservoir
is passed. I once saw a roe deer here which gracefully
cleared the wall to go cantering away over the fields. The
path comes out at the foot of the road's hairpin and the
outward route is rejoined. Note the beech hedge; and a
puzzle to occupy the remaining walk back to town: why
do beech hedges retain their leaves through winter while
beech trees shed theirs?

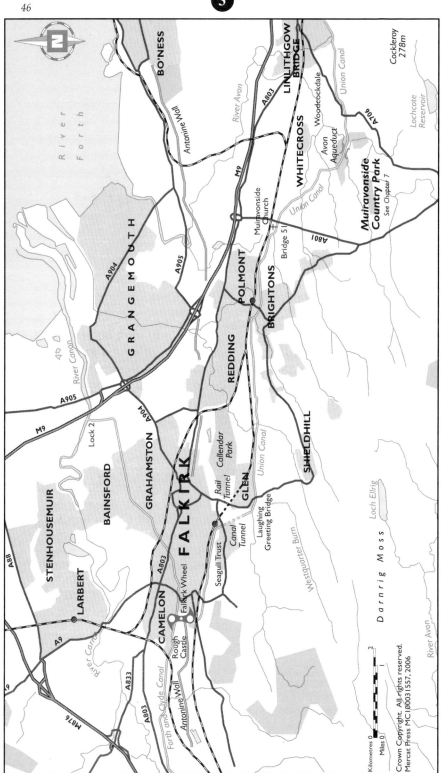

BO'NESS

LINLITHGOW BRIDGE

Cockleroy 278m

River Avon

A803

WHITECROSS

Woodcockdale

Union Canal

A706

Lochcote Reservoir

Antonine Wall

M9

Avon Aqueduct

River Forth

Union Canal

Muiravonside Church

Muiravonside Country Park
See Chapter 7

GRANGEMOUTH

A904

A905

POLMONT

Bridge 51

A801

River Canal

BRIGHTONS

A905

REDDING

SHIELDHILL

M9

Lock 2

A904

GRAHAMSTON

Callendar Park

Union Canal

Loch Ellrig

STENHOUSEMUIR

BAINSFORD

FALKIRK

Rail Tunnel

GLEN

Laughing Greeting Bridge

Darnrig Moss

A88

A803

Seagull Trust

Canal Tunnel

Westquarter Burn

LARBERT

CAMELON

Falkirk Wheel

A9

Rough Castle

Forth and Clyde Canal

Antonine Wall

A833

A803

River Carron

A9

M876

River Avon

Kilometres 0   1   2
Miles 0   1

7
LINLITHGOW TO MUIRAVONSIDE AND MUIRAVONSIDE COUNTRY PARK

OSLR 65; OSE 349

As one heads west from the Manse Basin good views open up to St Michael's church with its crown of thorns and the historic palace. Bridge 45 goes under Preston Road, where an early reconstruction was carried out to allow Linlithgow-based craft to reach the Avon Aqueduct. 1991 saw the culvert replaced by the present bridge. This also did away with the amusing but necessary red triangle warning sign on Preston Road depicting a swan. The culvert was too low even for a swan's height.

The golf course road passes under Bridge 46, a pipeline crosses the canal, then there's another underpass from derelict Kettlestoun Mains which leads to yet more quarries which supplied the stone to build Edinburgh's New Town. Over on the right is the Telford Avon railway viaduct with its 23 arches. The building of these viaducts owed much to the skills learned from constructing canal aqueducts, so there's a deal of irony in the railways killing off the canals. The canals across Scotland were eventually bought by railway companies which, irony on irony, kept them alive till the demands of road traffic forced the closure of the canals and the continuing parlous state of many railways, not, however, between Edinburgh and Glasgow.

During the canal's eclipse years when it was technically a 'remainder' waterway, a stretch of canal here was sealed

The magnificent Avon Aqueduct

off and empty because of constant leaking. The Kettlestoun
Breach was tackled pre-Millennium by monies raised both
locally and internationally, and I can recall the thrill of then
being able to cruise on one of LUCS boats out to the Avon
aqueduct.

Some cobbling leads to the next bridge, which has a
wrong (and confusing) 45 carved on the east side and
47, correctly, on the west. This is a much-repaired small
bridge, and the A706 now runs close beside the canal. On
the other side a concrete structure may be noticed. This is
a defunct canal water intake from a stream angling down
more or less parallel to the canal. When not needed, the
stream ran through below the canal, as a scramble down
the bank would show. Just ahead is Woodcockdale, one of
the old change houses where relays of horses for towing
would be swopped over. The human carriers had priority
over goods barges to the extent of having a rising, sharpened prow to cut through any towropes that got in the
way. Horsemen usually went ahead to clear the way less
dramatically. The building has been well preserved and is
used by Sea Cadets. The English-sounding Woodcockdale
name has been found on a map of 1491.

There is a car park just west of the Woodcockdale build-
ings reached by driving down and along the towpath.
Bridge 48 immediately after is a much-patched one, as the
A706 Lanark road gives it a hammering of heavy traffic.
Under the bridge, on the south side, there is another 'stop

gate', but as like as not these days truckloads of material are just dumped into the canal to form a barrier and the water pumped out.

There's a winding basin and ruined landing stage not far beyond Bridge 48. A swan's nest sits at its edge. All the young swans are ringed and studied. In winter they tend to gather on Linlithgow Loch or on the eastern seaboard anywhere between Montrose Basin and Northumberland.

The next stretch has a sad number of dead elm trees, the result of the lethal Dutch elm disease caused by a fungus which, once into a tree, stops the water rising and eventually kills it. The spores are transferred by clinging to the hairs of a specific beetle which then lays its eggs under the bark of elm trees, a cycle which has so far defeated all efforts at control.

Steps coming up bring the River Avon Heritage Trail onto the towpath. This trail, downstream, leads to Linlithgow Bridge, which offers another pleasant circular walk and, once across the Avon Aqueduct, continues upstream through Muiravonside Country Park—of which more shortly. Past mooring rings and a bend and you find yourself suddenly confronted by the canal's most spectacular engineering feature, the great Telford-inspired Avon Aqueduct. Only Telford's 1805 Pontcysyllte Aqueduct in Wales is larger in the whole of Britain. The 1823 *Companion* declares 'This noble edifice, which, for magnificence, is scarcely equalled in Europe, consists of twelve arches, is nearly 900 feet in length and 85 in height. The woody glens, the rugged heights, and the beautiful Alpine scenery around, must raise sensations of pleasure in every feeling heart'.

Downstream lies the Telford Avon railway viaduct, now a grade-A listed edifice. The Avon's waters drain from the Bathgate hills to the River Forth. On the river below here an old priory was largely washed away by one spate, so only a gable stands on the edge of the dell. In pre-glacial times the Avon flowed through what is now Linlithgow Loch to reach the sea at Blackness. Beyond the railway viaduct is

the site of the Battle of Linlithgow Bridge, 1526, when
the Earl of Lennox was killed after being captured in
an abortive attempt to rescue the young James V (born
in Linlithgow) from the clutches of the Earl of Angus, the
head of the notorious Douglases. The Avon was much used
for providing power and past centuries saw several paper
mills established. Walkers can look down on a weir when
crossing the aqueduct.

The aqueduct remains spectacular and, from below, is geo-
metrically graceful. It was only possible thanks to Telford's
ingenuity in using an iron trough to carry the water instead of
the usual puddled (kneaded) clay, which was much heavier
and just couldn't be carried on slender, practical arches.
(Baird was able to build these arches hollow, with struts.)
Once across on the north side, if you go down a little, you'll
see a grated opening which allows inspection access to the
interior of the viaduct. Inspectors can work across inside the
structure.

If pioneering techniques went into these canals, some of the
ships to use them were historic. Henry Bell's famous *Comet*,
the world's first practical seagoing steamship, was brought
through the Forth and Clyde Canal for her first overhaul at

The Avon Aqueduct, largest on the lowland canals

Bo'ness, where Bell had served his apprenticeship. The spectators ran from the harbour as she arrived. They assumed, from the smoke, that the ship was on fire and might blow up at any moment.

At the west end of the aqueduct the River Avon Heritage Trail heads off down a steep line of steps (over 100) to continue upstream through Muiravonside Country Park. As the park is well worth a visit, a circular walk is recommended, but a better route for this starts further along and is described then. There are plenty of interests along the canal now too.

Almost at once you come on an old milestone: 7 (to Falkirk), 24 (to Edinburgh). You'll notice how all the new milestones have kept to the same shape and script as the original, but will have an MM on the back. They were made by the apprentice masons of Historic Scotland. The stones were always angled so as to be easily read from passing traffic. The pillar next to the milestone is another of the four 'stage' posts, the last if heading west and inscribed as 'betwixt the third and fourth stages'—if you can make it out. On the other side of the canal there was once a dry dock. The ghostly hulk of the last remaining steel barge can be seen in the dock, but is so rusted at water level as to be beyond salvaging.

The dock worked very simply by just stopping the end next to the canal, and then releasing the water inside which drained down by a side stream into the Avon. The modern house that can be glimpsed sits on the site of an old canal cottage. Swans have continuously nested in the dock for many years. They eat floating duckweed which tends to accumulate in sheltered spots like this.

The canal is very rich in pond life, as can be imagined. Immediately before the next bridge there is another stretch with clear kicking stones. There's winding hole on the far bank.

Bridge 49 is a bigger bridge showing signs of remedial work and the steps up are no longer used. (Access to the road is 100m further on.) Shortly after, abutments on both banks point to a past railway crossing, and then (on the far side) you can look in to a large square basin, the Causewayend or Slamannan Basin. Coal was transferred here to barges, the lines jutting out over the basin, so, when the doors were opened, the coal fell straight into the holds. The basin has been restored and there are stagings/moorings. The basin can be visited on the Muirvavonside circular walk described shortly, but a quick visit can be made along a canalside footpath reached by a stile at Bridge 49.

Bridge 50 is soon reached (Almond Bridge) and there are public paths on both sides of the canal, as well as the towpath. I'd suggest crossing to follow the path on the south side, which wanders through attractive Scots pines with better views than the towpath—which can then be regained at Bridge 51, the completely rebuilt, stylish Vellore Road Bridge. The small Almond road not only spans the canal, but also the line of an old mineral line which crossed the canal, as abutments indicate. Hidden from the towpath, the Almond bing lies south of the canal, filling the area between Slamannan Basin and Bridge 50.

At Bridge 51, Muiravonside Church and Cemetery is signposted, but they can wait till the next section for our Muiravonside circular starts at Bridge 51 and is now described. It is signposted at the south end of the bridge, descends steps to picnic tables and, 100m on from the gate, at a marker post, breaks off right as a small path. Go through a gate and up a field edge, cross a small road leftwards and into another field (it is well signed). Go straight across this field to descend to a footbridge into a wood. The Almond bing is clearly seen to the left. Follow the path just inside the wood, coming to a cross-paths. Turning left leads along to the Slamannan Basin with the ash bing above—which can be climbed through the wood before the cross-paths; the flat top is laid out for bikers. A secretive spot. Pity about the litter. Return to the cross-paths to regain the route on: go through a stile/gate to follow another field edge (and house drive) to a stile onto another small road. The Muiravonside Country Park entrance is across, left.

Walk down the drive, crossing the Bowhouse Burn, ignore roads off left (to car parks) and reach an area with a confusion of paths, tracks and roads. The gorge of the River Avon is noticeable on the right and the major car park to the left. From the big car park it is worth walking the few minutes to visit Newparks Farm where a wide range of 'foot and fowl' is seen. (Behind the farm, down towards the Bowhouse Burn is a listed limekiln.) Return to the car park and take the road on to the Visitor Centre which occupies the stables and outbuildings of the one-time mansion of the estate. There's a small tearoom, but its hours are erratic! The Visitor Centre has exhibits and explains much about the park. Pick up a map before leaving

Doocot in Muiravonside Country Park

There's a play area next to the Visitor Centre and a rather fine 'lectern' doocot. At Linlithgow we saw an earlier style, a 'beehive' doocot; these eventually gave way to rectangular buildings with a single sloping roof, imaginatively compared to a lectern in a church. The one here is a link with the third style which were purely ornamental often as part of stable blocks or set on some eye-catching spot, really a folly rather than supplying any need for pigeons as a food source.

Behind the tearoom the outline of the mansion can be seen, and there's a signpost of the River Avon Heritage Trail with its blue markings. From here several paths head off and one can zigzag down to follow closely by the river or take a higher line. For now, I'll just mention one variation, following this higher line. In early 2006 there were some path closures as the old mill was being restored, but if one strays, despite my suggestion, it can only be temporary: the Avon Gorge is always on the right and has to lead to the great canal aqueduct.

Along the higher path at one point a few steps go down to a cave-like grotto, typical of its kind. There's even a fire-place. Further on there's a group of three perching seats and, on again, a small walled graveyard of the Victorian Stirling family who once owned the Muiravonside estate. David Stirling of SAS fame was a more recent member of the family. (A striking statue of him stands looking to the Highlands on the Bridge of Allan-Doune road.)

The path dips and forks. Keep left (right was closed in 2006), along above what was once a riverside garden's site with a long pond (a curling pond?) and mill lade just visible. Steps lead up onto open ground. A path off right just before this exit drops down over the lade to the mill—and would be quickest, if open by the time of reading this, otherwise go along outside the trees until able to descend on a track to the riverside, where there's another big signpost.

Turn left along towards a fork and take the right, riverside option; this passes an arched passage below the left track, passes another group of the perched seats and crosses the Bowhouse Burn by a substantial wooden bridge, which becomes a steep, twisting flight of steps up and under the canal aqueduct. The views of the aqueduct are best in winter when the trees are bare. Quite how many steps up there are is a fruitful source of disagreement for any group: a bit over the hundred anyway. A surprisingly quick tramp along the towpath leads back to Bridge 51.

8
MUIRAVONSIDE TO THE FALKIRK WHEEL

OSLR 65; OSE 349

Head off along the towpath from Bridge 51. Shortly there is an overspill across the towpath with steps up and along by a burn to allow access to Muiravonside Church. There's a large parking area. Beside the church the old walled graveyard has a superb collection of 18th century stones, many obviously by the same masons, craftsmen with a taste for fanciful supporting trumpeters, symbolic circles and other emblems of mortality and immortality—simply one of the best collections of this folk art which neglect is steadily destroying.

Continue, once back at the overspill, to Kirk Bridge (No 52), which simply leads over the canal to a field but is a good

Trumpeters of the Resurrection on one of the fascinating old stones in the Muiravonside churchyard.

Ben Vorlich and Stuc a' Chroin from the Muiravonside Kirk Bridge

viewpoint for Stirling Castle, the Ochils and Saline Hills (pronounced 'sal-in'), and with the familiar shapes of Ben Vorlich and Stuc a' Chroin visible. The extensive Manuel Works (refractories, terracotta products, and much else) lie beyond the graveyards and, in the middle of this sprawl, is the square tower of Almond or Haining Castle, once a seat of the Earls of Linlithgow and Callendar, but derelict for over 200 years.

Rural walking leads on to a bridge which caused some problems of restoration. Its number 52A indicates there wasn't a bridge here originally, so when the A801 was built to link M9 with Bathgate and Livingstone the canal was just culverted. The problem was similar to the M8 nearing Broxburn, as here too the river puts in an S bend. The A801 runs uphill, and on the original canal line there would not have been enough headroom to pass under the road, so the canal bent left and then right to pass under where there was headroom. Neat.

Bethankie Bridge follows—or aqueduct, rather, as this time the canal does the bridging, and the feature is easily overlooked as there is just a short curve of parapet. There's arch on arch, for while the canal spans the road, the road itself spans a burn!

The Bethankie Aqueduct where the road goes under the canal

There is a clear view across to the vast array of cooling towers and so on that mark Grangemouth, but they soon drop out of sight as you walk this open stretch, the railway in a cutting beside you and new Polmont houses ahead. Despite the feeling of suburbia, only a farm track crosses Bridge 53, and you hardly see Polmont thereafter as you enter a hemmed-in cutting leading to the large span of Bridge 54. An iron pedestrian bridge has been added on the west side. There's a shop and other facilities near at hand, and also, Polmont station which, with others, makes walking stretches of canal easy as a return to the start can be made by train. Polmont to Falkirk High or Linlithgow are prime examples.

From the Polmont bridge there's access on both sides of the canal briefly, a winding hole and an overspill which goes into the Polmont Burn which flows under the canal. It joins the large Westquarter Burn to head through Grangemouth into the sea at the docks. A smaller bridge with blue railings links new housing with the town. The canal swings westwards with the view to the Ochils opening up. Across the canal are the grounds of the Grange Education Centre where the Waterways Trust Scotland / Falkirk Council new

'Action Outdoors' centre is based, offering such activities as barge navigation, canoeing, cycling, climbing, walking, fishing, hence the new moorings etc. Showers and changing facilities are available.

The next bridge is a non-standard concrete one which forces pedestrians on an 'up, over and down'. North there are railway sidings and the line from Polmont and Edinburgh divides, one branch heading to Stirling (via Falkirk Grahamstown) and the other, the canal-follower, heading to Glasgow (via Falkirk High). Redding number 23 pit beside the canal was to be the scene of one of the worst mining disasters in Scotland on 25 September 1923. Water from abandoned workings burst into the pit and 40 died. Five miners were brought out alive after nine days, and it was nearly three weeks before rescue attempts were called off. In Old Polmont graveyard I found a heartbreaking stone that read, 'Colin Maxwell, aged 58 years. Also his two sons, Walter, aged 28 years and Colin, aged 17 years, who all died in the Redding Pit disaster …'

There's a staging, the canal is wide and runs along facing a well-fenced prison: a young offenders institute. Some sources suggest Burke was a navvy in the Polmont area, others that he was working on the Falkirk tunnel where he met Hare, and the two later plied their nefarious trade in Edinburgh.

Bridge 56 carries pipes across, and the whole south side (Redding Park) is undergoing major development under the title of Canal Corridor, with both new homes and commercial projects under way—so will change greatly in the life span of this guide. We join a section of road alongside a small north bank industrial estate. There's a winding hole and later a swing bridge abandoned on the far side, which once gave access to chemical works. As the only swing bridge on the Union Canal it deserves preservation. A travellers' site lies below the north side of the towpath. Bridge 57 also carries pipes on brick pillars, and passing under the canal is the large Westquarter Burn which so bisects Grangemouth.

A more rural feel returns, though the white geometric houses of Hallglen are starkly visible beyond the busy railway line which runs just below the canal awhile. Bridges 58 and 59 just lead to fields, the banks are wooded again, the canal wide and with possible winding holes. A cobbled overspill (through a wall pierced with square holes) falls into the deep set Glen Burn which runs into the Westquarter Burn. Shortly, curved walls show where the burn flows from under the canal. A footpath, off right, just leads to houses, then the big Bridge 60 spans our deep, wooded walk, one of the biggest arches over the Union Canal. What isn't noted from the towpath is that the Glen Burn runs in a deep trench above and parallel with the canal. Bridge 61, Glen Bridge, has a second span, with neat curving parapets under which the burn flows. Invisible from the canal, it is easy to walk up and see.

Another feature captures the attention: there are carved faces on both keystones, that on the east laughing, that on the west glum and girning, and popularly known as the 'Laughin and Girning Bridges'. Accounts vary as to their meaning. Did one smile at the long miles built from the capital? Did the other grimace at the work just ahead in creating what was probably the first-ever transport tunnel in Scotland? It has also been suggested the masons showed their feelings for the characters of their respective bosses when

Laughing and Girning faces on Bridge 61

Kicking stones by the
Falkirk Tunnel

the work met here. Above the faces are ovals with the number 61 and the date 1821. Along the 100 yards leading up to the tunnel are excellent examples of kicking stones.

There was really no need for a tunnel, but the 18th century industrialist William Forbes, who had bought Callendar House (forfeited from the Jacobite Livingstones) objected to the proposed route, as it would be in view from the palatial chateau he was making out of the old house. The building and its parklands now belong to Falkirk Council and there's a good stretch of the Antonine Wall's Vallum visible in the park.

The 690 yard (631m) tunnel through Prospect Hill was quite a construction feat when you realise it was cut by navvies working with horses at best—and none of today's power tools. Three shafts were sunk so the tunnel could have several work faces operating at once. It was 18 feet (5.5m) wide (13ft of waterway, 5ft of towpath) and rather more in height (6½ft of water, 12ft clearance, total 19ft/6m). The tunnel is lit, but still has a somewhat spooky atmosphere. There is a sturdy handrail for safety. Where water runs down the walls or drips from the roof some beautiful formations have been created. Just to the north the Edinburgh-Glasgow railway also runs through a tunnel.

You come out from the tunnel to a park-like area, with tidy grass verges and well-made paths, Falkirk's vast sprawl half-glimpsed and a view to the distant Ochils. Falkirk High station is just off to the right. There is a basin before the Bantaskine or Walker's Bridge (number 62—which is the last), and the towpath is even tarred. A nice moment comes when the screening on the right finishes and there is a clear view of the Campsies. Ben Ledi lies to the right of their sprawl. The Seagull Trust's *Govan Seagull* (built by Govan shipbuilders in 1984) operates from a reception centre in Bantaskine Park, allowing disabled visitors to sail through the tunnel or out to the Falkirk Wheel. The *Barr Seagull*, built in Falkirk, was added to the Seagull Trust fleet in 2006. The woodland park has attractive paths and a quiet ambience.

The canal continues along above Falkirk (its Steeple clearly visible) to a milestone with '½ mile' on it—indicating the original distance to the finish of the canal. Beyond, the incut on the right marks a historic spot, though there is virtually nothing to see today except a dirt track running down to the road that goes under the massive arches of the railway viaduct. Having dutifully kept to its 240 foot contour line all the way from Edinburgh, the Union Canal ended 112 feet above the Forth & Clyde Canal so, perforce, they had to be linked by a flight of locks, eleven of them, leading down this line to a large basin, Port Downie

Port Downie and the Union Inn, where the Union (left) and Forth & Clyde Canals met. (©East Dumbartonshire Council, Wm. Patrick Library Kirkintilloch)

(now gone—named from an investor, Robert Downie of Appin)— beside Lock 16 of the Forth & Clyde Canal. (Lock 16 is wrongly shown on OSE 349.) The Union Inn took its name from this historical spot, and at least its welcoming presence still survives. Old photos show the locks swinging under a six-arched railway viaduct but, at the time (1961) of filling in the locks and creating roads, this was replaced by the big arch we see today.

Let me indulge in a personal note here. I was once trying to photograph the Union Inn and the locks on a showery day and was constantly in and out of my camper van during the mini-deluges. With a variety of doors the inevitable happened: I found I'd locked myself out. After some thought I went in to the inn and, all too audibly, asked the barman for the loan of a bar stool, a broom and a wire coat hanger. This caused a speculative hush and when I carried out my collection of objects I was followed by most of the drinkers, obviously curious to see what was afoot. Well, the stool let me climb onto the Transit's roof where I lifted open a skylight through which I was able to insert the broom, at the end of which I had fashioned a hook from the wire coat hanger. Very carefully I hooked the bunch of keys lying on the sink draining board and drew them through the skylight. Success was greeted by a polite round of applause.

Alternative plans to connect Union Canal with the Forth & Clyde Canal, mooted by Robert Stevenson and Thomas Telford, were to carry it on to the Wyndford lock, but several locks would still be needed and another big aqueduct across the Castlecary 'gap', aswell as cutting more miles of canal. There's an element of sadness in the disappearance of the splendid flight down to Port Downie which was made.

This, of course, set a big problem over linking the two canals again. Reinstating the original line was no longer practical: roads, buildings and a town's infrastructure ruled that out. So what was to be done? The answer was bold and ingenious, a truly brilliant concept which came to fulfilment in the instant icon of the Falkirk Wheel: a rotary boat lift which would simply pick up a boat from the higher level and gently lower it to the basin below (or vice versa). To make room for this the Union Canal was extended, so, after a look at the spot where the flight broke off, we can push on and reach the unique spectacle of the Falkirk Wheel, afoot, an approach only to be improved on by being on the canal itself.

Continue then through the gate and on to a Y-junction of the canal: the short section left, often with B W Scotland working barges present, is the old ending of the Union Canal (an 1823 extension, Port Maxwell, from which passengers walked down to Port Downie). The right fork, the new continuation, is another MM creation, Summerford Aqueduct, which strides across a road twisting up underneath (the road climbs up to the monument on the site of a 1746 battle, of which more later) and heads on westwards. An old bridge, used as a footbridge, crosses the railway cutting running parallel with the canal, leading into the Tamfourhill housing estate which has to be passed before the linking begins. A dirt track then crosses both canal and railway at a basin, where the walker has to do another up-down road crossing, to then follow a wider towpath suitable for maintenance vehicles. This is the ending of the Union Canal today.

There is staging, then two locks lead down to a large basin. A plaque on the building at the locks commemorates a local Thomas Douglas who immigrated to Canada, became a Baptist minister, then entered politics to become premier of Saskatchewan. The plaque points to an attractive commemorative seat by the north portal of the tunnel. It was erected by the Camelon Local History Group with aid of HMYOI.

The locks drop down to the large basin, then the canal swings northwards, vanishing into the 180 metre Rough Castle tunnel, which passes under the Edinburgh-Glasgow railway, road (Camelon-High Bonnybridge) and the line of the Roman Antonine Wall, to come out right onto the Falkirk Wheel aqueduct—a great effect. Pedestrians generally descend, off right, no doubt to make a bee-line to the Visitor Centre and book a trip on the winged wonder, but a neater approach is to take the serpentine path up and over the top of the tunnel entrance and down the other side, passing under the encircled meeting of aqueduct and Wheel to reach the Visitor Centre.

The striking image of the Falkirk Wheel is almost as well known as the Forth Bridge (a deal younger of course!), yet the practical operating of the system is basically simple. There are two balancing gondolas (one with boat or boats in it) which are set in motion so that as one descends the other goes up. So fine is this balance that the electricity needed costs less than £20 a day. But the joy of this creation is the marriage of engineering with what can only be described as art, architecture as sculpture, practical, inspired, Archimedes for today.

The Visitor Centre itself is of bold design, and has various displays, café and gift shop and sells a range of books about canals, including those in print about these Lowland canals. There's a working model of the Wheel and an audio visual on the Millennium Link that brought the canals back to life.

The large basin *cum* marina, New Port Downie, has all the facilities for canal users and is joined to the Forth & Clyde Canal by the Jubilee Lock. Pity about the inappropriately visual childrens' 'Fun Factory'. An excellent 1-2 hour circular walk can be made from the Wheel to visit Rough Castle, the Roman Fort, and this, with other options, is described in Chapter 12.

A signposted path leads to a rotating footbridge across the Forth & Clyde Canal to make the final walkers' link of the two canals. What next? The purist must explore the stretch down to the Forth Estuary, and this is described at the start of the next chapter which then describes something of Falkirk itself.

On the Falkirk Wheel Aqueduct

The Falkirk Wheel

9
THE DESCENT TO THE FORTH

OSLR 65; OSE 349

The Wheel is neatly the midpoint of the joint canals with Bowling lying 32 miles west and Edinburgh 33 miles east. The Forth & Clyde towpath is joined over the new swing bridge. The south bank moorings are the base for many of the commercial longboat hire companies. The *Marion Seagull* (2003) is the first Seagull Trust boat with family/group accommodation, allowing disabled cruising (2-6 nights). A mile of walking east leads to Lock 16 and the Union Inn. The sprawl of Camelon (no Camelot, and pronounced Kam-lin) lies on the left. At Lock 16 the Carron Sea Lock is indicated as three miles on.

The name Union Inn harks back to the canals in their prime, for this was where the Union Canal descended in a stairway of locks to unite with the Forth & Clyde Canals, at the large basin of Port Downie, busy with traffic. The basin and flight of locks were closed in 1933 and filled in thereafter. The attractive Georgian Union Inn is still a welcome hostelry, (with canal murals and photographs), as is the Canal Inn by Lock 16, which lays claim to be the oldest canal-side tavern on the Scottish canals.

In 1839 an experiment was carried out with canal boats towed by steam engines running along towpath tracks, a length of track was laid by Loch 16 and various weights of boat were moved, all very speedily and successfully. Applying the system throughout would have been prohibitively expensive, so the project was abandoned. (The Panama Canal would see just such a scheme used.) Iron and chemical industries

The Canal Inn at Lock 16

once existed here and, more unusual, a nail-making works. All tended to have chimney stacks belching out a pollution mercifully absent today.

Lock 16 starts a flight which drops steeply to Lock 7 and then ever more gently towards the sea. There is now nothing to see of the one-time flight up to the Union Canal. The near-impossibility of resurrecting such a link happily led to the brilliant concept of the Falkirk Wheel.

The footpath undulates down beside the locks (there's a footbridge over by Lock 14) with Lock 11 introducing another interesting spot. The Rosebank Distillery buildings dominate, with the Rosebank Beefeater Restaurant/Pub on the left as Lock 11 is reached. The remaining distillery buildings, with the chimney stack, lie across the busy road junction. The Beefeater is in the former bonded warehouse, and the interior is an impressive conversion and worth seeing. The other buildings are due for redevelopment, and flats are already going up overlooking the canal as the descent continues. Lock 11 had to be re-sited to allow boats to pass under Camelon Bridge. Originally many bridges used the sturdy structure of locks as foundations, but their simple lifting or swing styles were inadequate for today's heavy traffic. Both Lock 11 and

5 were moved to solve this road problem, the alternative being to raise the road levels—not really practical. Look out for a lock-keeper's cottage, Lock 9, before a railway crossing forces walkers up, round, and down again. (The railway was originally a swing bridge.) Lock 8 leads into the busy Dollar Industrial Estate, a modest echo of the iron and chemical industries once based alongside the canal. St Mungo's High School lies on the left, an architectural nonentity.

Staging by Lock 5 lies beside the Boathouse Bar and Grill, another conversion from industrial use, and the canal drops below Bainsford Bridge for a longer straight, with an industrial estate on the left and some housing on the right. Beyond Bridge 4 (Abbotshaugh) there's a more rural feel. After a modern bridge (Orchardhall; painted light blue) there's a large overspill (over on the right are ponds and marshes) and then an abrupt turn to the left.

The original line of the canal went straight ahead here but re-instatement was never possible. The original line can't even be walked with the frantic M9 just ahead. Clear vestiges of the line can be seen off the Skinflats A905 near the Kerse Bridge over the River Carron and from Kerse Bridge it is possible to follow a path up and under the M9 to look across to where the Forth & Clyde Canal enters the River Carron. A track leads down to the foreshore in the other direction. None of these interesting places can be reached from the canal at present. The canal, unable to follow the original line down through Grangemouth, was ingeniously turned here and runs along to a new Lock 3 (the original was at the bend) where one should cross. From the bend to Lock 2, where the river is reached, is the Carron Cut, virtually the line taken by a private cut for the Carron Ironworks which was never operational.

The end/start of the canal has the feel of a small, slightly down-at-heel marina, with machinery and materials lying around, port-a-cabin offices, etc. A welcome number of pleasure boats are visible, however, and, despite, or because of the M9, the spot has a restful air. Finding it by car (see p.70) is a navigational challenge!

The sea lock into the tidal River Carron is Lock 2, and if this appears strange numbering, there are plans for another lock on the river to improve access—a nice project for some rich benefactor! The large stumpy chimney in sight over the last couple of miles rises across the Forth Estuary: Longannet Power Station (pronounced: Long-annat).

Grangemouth was established when the canals were built, with docks and industry clustering its length. (The original route went through the docks.) Grangemouth has fared better than other canal-based towns, with modern oil refineries and related industries giving the place a lit-up surreal appearance at night. The local Heritage Trust maintains a museum giving the town's story. One claim to fame was that in 1803 the first practical steam-powered vessel, the *Charlotte Dundas*, was built here and underwent trials up and down the canal.

The name Carron, from 1778 into the 19th century, was famed at sea for producing the Carronade, a short, hefty gun which was deadly at close range when it could fire 68 pounds weight of grapeshot, with deadly effect on rigging and crew. Nelson had two on the *Victory* at Trafalgar. The 'Smasher', as it was nicknamed, fell out of use with warfare being conducted at longer range. The works closed in the 1980s. Carron 'ranges' were once common in Scottish kitchens.

Carron Basin—the River Forth end of the Canal

Being now at the Forth's tidal waters, it is time to head for the waters of the Clyde. Unless one has a tame driver for a pick-up at Lock 2 (see below) there is not much alternative to the hour's walk back up. Walking back up to Lock 16 and the Falkirk Wheel however gives a feel of continuity, which will be rewarded when Bowling is reached at the other end of the Forth & Clyde Canal: one has gone truly end to end, estuary to estuary, coast to coast.

At Lock 16 there is an alternative to following the towpath back to the Falkirk Wheel, and that is to walk along the attractive Watling Lodge section of Roman Wall between the two, or taking in Rough Castle as well. This is described in Section 10.

Now, to find the Carron Cut and Sea Lock by car. (OSE 349 the best map.) If crossing Kincardine Bridge, join the M9 (Edinburgh direction), and/or if on it already, continue, to cross the River Carron and turn off at Junction 6 slip road. Turn right and then at the big roundabout with the motorway flying over it, take the A904 Falkirk exit. If coming from the east (Edinburgh) turn off at junction 5 (Grangemouth) onto a roundabout with motorway over. Pass under the M9 and turn first left, the A905, which is followed through the next roundabout to the next motorway-style roundabout, pass under the M9 and take the first left, A904.

This is an awkward bit of dual carriageway with the West Mains industrial park on the right. Turn into it (first right) and along past the Asda building. The road swings right and 150m on, left, is the signed and gated approach road to the Carron Cut/Lock 2. (The gate is locked 8pm-8am in summer, 4.30pm-8am in winter.)

If in Falkirk, find your way to Camelon and the canal. Join the A9 and stick to it. The road snakes clear of Bankside to swing right and over the canal (there are blue railings on the bridge). Keep straight on at the first roundabout and left at the next which is the A904. Turn left into West Mains industrial estate as indicated.

Falkirk Town

The slopes above UC Bridge 62 saw the last Jacobite success of the Forty-Five rising just three months before Culloden. There is nothing to see on the ground, and the roadside monument (867789) is a crude concrete obelisk. The story is not without interest though. Lord George Murray surprised Hawley, who was camped at Falkirk, and drove his force back to Edinburgh. The clash occurred on a windy, sleety day in January 1746. The Livingstones, Earls of Linlithgow, had Jacobite leanings and were forfeited after the Fifteen, but the Falkirk 'Bairns' (as the town folk are still called) refused to pay rent to the York Building Company and the estate was leased back to the Countess of Kilmarnock, the Livingstone's heiress, whose husband had 'come out' for Prince Charles, was captured and beheaded. Ironically, Hawley, leader of the government forces, was dining with the countess when the Jacobites attacked his forces.

Callendar House saw most of the regular figures of note: Mary Queen of Scots, Cromwell (it was Monck's Scottish HQ) and Charles Edward Stuart, en route for Derby. One of the unenthusiastic participants in the battle was the Gaelic poet Duncan Ban MacIntyre, who is forever linked with Ben Dorain. An earlier Livingstone was principal guardian of Mary Queen of Scots, and the nobles at Callendar House in Falkirk had to decide if Mary and Edward, son of Henry VIII, should be betrothed. They decided no, and Mary went off to France for safety. One of her Four Maries was Mary Livingstone.

The first bloody battle of Falkirk was back in the time of the Wars of Independence, and was the sad end to Wallace's efforts to free Scotland from English interference. He had finally succeeded in driving out all the English garrisons, had been appointed 'Guardian of Scotland' and carried fire and sword into Northern England. Edward I was in Flanders fighting the French king, but was forced to return and invade Scotland in 1298. Wallace's smaller, less well-trained force was caught at Falkirk and, despite bloody resistance, was simply massacred by the sheer weight of English numbers and the deadly

longbow which weakened the 'schiltrons' of fierce spearsmen.
Wallace continued the struggle, tried in vain for continental
support and, on his return, was betrayed. Edward had him
barbarously hung, drawn and quartered as a 'traitor', which
ensured he has been revered ever since.

This central corridor of the country was much fought
over, for Stirling was the lowest bridging point on the Forth
and so forced communications in that direction, for good or
ill. We seldom remember how greatly history is affected by
geography, both in the big events and battles and in everyday
social activity and trade. Falkirk as the 'epicentre of Scotland'
could hardly escape. Now it is a town of 40,000 inhabitants
and a busy place, despite the decline of older industries. Coal
mines and iron foundries made it a leader in the Industrial
Revolution.

It was an East Lothian entrepreneur who began this industrial revolution, but his local mine-owner would not reduce
prices to make production viable. The famous Abyssinian
traveller, Bruce of Kinnaird, had pits near Falkirk and he de-
livered the goods. So Carron, now swallowed up by Falkirk,
became the heavy iron industry centre of Scotland. Roman
camps, town walls and old buildings have all gone ,but many
18th century buildings are now carefully preserved, there is
a Town Trail for pedestrians (and a longer Town circuit for
motorists), an excellent museum and the rebuilt town centre,
of huge glass and concrete modernity, is well surrounded by
parks and gardens.

Of interest to walkers were the annual Falkirk Trysts, the
largest cattle marts in the country. The scale was enormous:
60,000 cattle and 100,000 sheep are said to have been sold in
one day. The drovers' routes, from the remotest Highlands, are
one of our treasured legacies. The Highlands were thickly peo-
pled then of course, but this does show a level of population
and production that could be obtained again,were the powers
that be genuinely interested in the Highland economy, which
stutters along on a feudal system of land ownership damag-
ing to past history and future hopes. Beasts could not be fed

in winter, so the great trysts saw the surplus sold off. Many were walked on to the industrial cities of England, or even to Smithfield Market in London. The tryst was held to the south-east of Falkirk and later moved to Rough Castle, and finally to Stenhousemuir, whose earlier name was Sheeplees.

The easiest way down to the town centre (or for Callendar Park/House) is just after coming out of the canal tunnel, which also gives access to Falkirk High railway station. From the station head down High Station Road which merges with the B8028 at a rather complex junction. Continue down for the town, turn up and first left (Kemper Avenue) for Callendar Park/House. Heading down, Comely Park primary school is passed, keep on ahead at a junction and, as the road swings right, take St Crispin's Place (Cow Wynd) left which leads to the pedestrianised High Street.

Despite being full of identikit multinationals the centre has a lively air. Turn left towards the Steeple. The narrow lane before it is called Wooer Street; while other odd names are Bean Road, Ladysmill, Tanners Road and the Tattie Kirk. The Steeple, dominating the High Street, houses the Tourist Information Centre which can help with accommodation, town maps, leaflets, etc. This is actually the third Tolbooth Steeple. The original was rebuilt in 1697, but a century later the demolition of adjoining property so undermined the foundations it had to be taken down as well. For eleven years there was no Steeple, but funds were raised to build the present 140ft (43m) spire. The top section was rebuilt in

The Steeple, Falkirk

1927 after suffering a lightning strike. The weather cock went flying and masonry crashed everywhere, but the only fatality was a horse belonging to Mr (Irn-Bru) Barr. The Cross Well beside the Steeple dates to 1817, replacing one originally given to the town by the Livingstones of Callendar in 1681. The site of the Mercat Cross (and of the town's last public hanging in 1826) is marked out on the setts (rectangular cobbles). Tolbooth Street, behind the Steeple, makes the *Guinness Book of Records* as the shortest street in Britain.

Continuing, turn in through an arch to reach the solid Old Parish Church with its octagonal tower, and the site of one of the oldest historical tombs in Scotland, that of a Graeme killed at the Battle of Falkirk in 1298. The church was rebuilt in 1771, and again in 1860 when the arched crown of Gothic ironwork was added. The church dates to 1810, though the tower is earlier, and the site goes back to the start of historical time. The grounds were cleared of gravestones in 1962, except for a few historical ones, like Sir John de Graeme's. There are tombs to victims of the 1746 Battle of Falkirk too: William Edmonstone and Munro of Foulis and his doctor brother. Foulis must have been a paragon; 'His death was universally regretted. Even by those who slew him'. Good value lunches are served daily in the church hall.

There are some impressive civic buildings; showing good Victorian confidence as so often. The Jacobean sheriff court was completed in 1868, but when the first Sheriff was appointed in 1834 everyone, perforce, had to use the ballroom of the Red Lion. The Carnegie Library is Gothic with some fine glass in the windows. Walking out westwards, the Forth & Clyde Canal can be joined at Rosebank/Camelon, on the way passing, on the right, the attractive Dollar Park, named after Robert Dollar, who left it to the town. He was a Falkirk Bairn who emigrated to Canada and made a fortune. There are plenty of flowers and mature trees, and the large glasshouses produce nearly a quarter of a million bedding plants each year for use in the district. Beside the pavement stands the war memorial. A plaque for the first

World War notes the horrific figure 'Over eleven hundred Falkirk Bairns died'.

The most interesting feature of Falkirk, for walkers especially, is Callendar Park and Callendar House, ten minutes walk east from the High Street. Aiming for it from Falkirk High Station has been mentioned; about 300 yards along Kemper Avenue there is a road into the magnificent Callendar Wood which can be circuited or, if Kemper Avenue is followed, a path off to the mansion once past the trees. Where to start? At Callendar House one can pick up leaflets on the many attractions.

Within the house there are exhibitions, a really excellent historical 'walk-through', a Historic Research Centre, and living experiences of a Georgian kitchen, the interpreters in period costume. You can glimpse the work of a clockmaker, printmaker and grocer too. Five minutes away, in the old stables block, there is an art gallery of contemporary work and an attractive tearoom. Walking straight out from the entrance of Callendar House, an ice house is passed before the road cuts through the very clear line of the *vallum* (ditch) of the Antonine Wall. This bit of vandalism was done to make a grand entry for a visit from Queen Victoria; but her visit was cancelled. East of Callendar House a lime tree avenue leads to a large man-made loch (for boating etc) while to the south lies an arboretum of mature trees and the slope of the great wood, offering endless explorations. The loch once fed a canal, which accounts for what looks like a ha-ha between house and woods. There's a range of amusements for children and a pitch and putt course. However, the park-like setting, with the château-like mansion backed by the great wood, is in itself a delight. A place to visit often. Chapter 10 describes the Roman sites of the area which offer attractive rural walking, then Chapter 11 continues the canal description westwards, on to Auchinstarry.

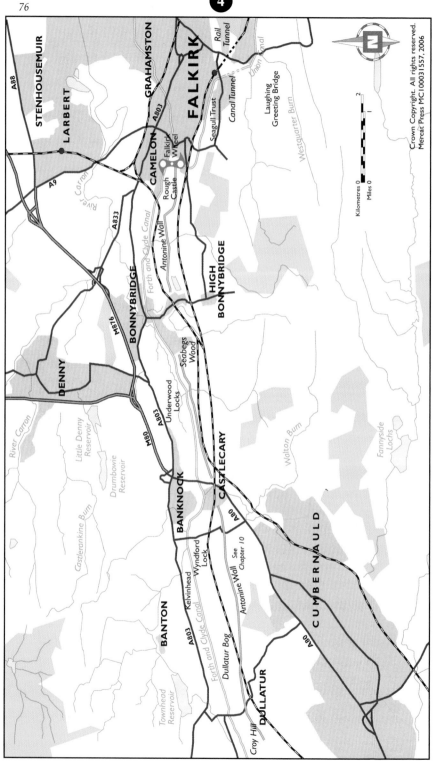

10
THE ROMAN WALL: WATLING LODGE, ROUGH CASTLE, SEABEGS WOOD

OSLR 65; OSE 349

West of Camelon (Lock 16) and running on to Twechar (west of Kilsyth) are the best-preserved sections of the Antonine Wall which, running more or less parallel with the canal, offer several side trips or excellent circular walks. Keeping these descriptions apart from those of the canal, Chapter 12 describes the sites over Croy Hill and Bar Hill; here we have the Falkirk area attractions and Chapter 11 follows the canal from the Falkirk Wheel on to Auchinstarry.

This great Roman monument deserves to be better known, though the feature has had a rough passage historically, with roads, railways, canals, buildings, industry and agriculture all wrecking its course across 'the waist of Scotland'. Perhaps the worst damage was done in the 18th century. Before then several of the sites had extensive ruins, by the end of that century the 'convenient' stonework had been carried off to build houses, field walls, canal banks and much else. An interest in historical remains came a generation too late.

The Wall runs for 37 miles (60km), linking the River Forth near Bo'ness to the River Clyde at Old Kilpatrick and generally commanding low ground facing north, with forts at regular intervals along its length. Some, like Rough Castle and Bar Hill, are still of interest. The Wall was built of turf, though on a stone base, so it has not survived very well, nor has what could be called the service road which ran along behind the Wall, though this Military Way, as with Dere

Street, is often indicated by the pockmarkings of small quarries alongside.

The best surviving feature has been the *vallum*, or ditch, which ran along in front of the Wall—and the best lengths of that feature will be described. The main road north beyond the Wall started at Camelon, just west of Falkirk, and there are traces of Roman forts and camps angling across behind the Ochils to Perth and up to the north-east of Scotland, far beyond Hadrian's Wall which tends to be regarded as the Roman's northern boundary. The boundary was seldom static in reality.

Caesar had first raided England in 55BC, and the real invasion began in AD4. The governor Agricola first entered Scotland in AD79, with the optimistic hope of adding it to the Roman Empire. Trimontium (Newstead) on the Tweed, named for the triple-peaked Eildon Hills, became the main Borders base and, along the narrow waist of Scotland, Agricola built a line of forts, quelled Galloway and headed up to the north-east to win the great battle of Mons Graupius, the site of which remains tantalisingly unknown. His fleet reached Orkney before he was recalled to Rome. Early the next century Rome withdrew from Scotland back onto the Tyne-Solway line, and a decade later the Emperor Hadrian decided a definite defence line was needed, and commanded the building of the wall that bears his name.

His successor a decade later decided to re-invade Scotland, which was undertaken in AD140 by the governor Urbicus, and so, a couple of years later, we had Antonine's Wall which followed the line of Agricola's forts. Indeed, several of the new forts lie on top of the earlier defences. The Wall was held for a decade, then about AD165, Antonine's Wall was finally abandoned for good. Three legions, about 8000 men, were stationed on the Wall. Rome's waxing and waning fortunes led to their eventual departure from British shores, but that is another story. In this section, three sites within walking distance of the canal are described.

Watling Lodge and Rough Castle

Watling Lodge offers one of the finest sections of Roman ditch on the Wall, still showing the deep V nature of this defensive feature. (Elsewhere the ditch is often much filled in and can be boggy.) With a wall on the south side of the ditch, we can envisage just how formidable a barrier the Romans created. The soil dug out was always heaped up on the north side. Walking along by Watling Lodge is well recommended, but if not wanting to, just repeat Lock 16 to Wheel by the towpath—and the walk can be extended to include Rough Castle, an impressive fort site.

Head south from the Union Inn (at the sign for Union Canal) and turn right on reaching the road: Glenfuir Road—which actually runs along the lower sweep of the locks originally linking the two canals. Turn first right onto Tamfourhill road. Cross it at once to a gate with a Historic Scotland sign for Watling Lodge. Walk along the north bank. All too soon this beech-lined ditch is interrupted: a brick building sits across it, one of the outbuildings of Watling Lodge. Sadly, the building appears to have been built on the main gateway/ road through the Antonine Wall which led to Camelon, site of Roman forts and a Pictish town brutally levelled by Kenneth MacAlpin as he forged a single nation out of the Picts and Scots tribes. You perforce turn right, back down steps (there's a good interpretive board) to return to Tamfourhill Road. Turn

The vallum *of the Antonine Wall at Watling Lodge, Camelon.*

left. Past the buildings the ditch can be seen again, soon
angling down to cross Tamfourhill Road, where a couple
of sections of ditch are signposted, but are discontinuous
thanks to other buildings—so keep to the road and at the
Tamfourhill roundabout turn right, on Lime Road, which
is the main drive in to the Falkirk Wheel which, of course,
could be taken, but there is a finer way to the Wheel a little
further along.

A pleasant, woody continuation is signposted on the
left of the drive, and at one stage passes over the northern
portal of the canal tunnel, with a fascinatingly geometric
view along the aqueduct to the Wheel with the Ochils hori-
zoned to the north. The views are even better when a gap
is reached. This is the better way back to the Wheel, enjoying a view few visitors coming by car would discover. The
sprawl of Camelon, Larbert and Stenhousemuir are backed
by the haughs of the Forth and the impressive Ochils. Paths
lead along and down under the Wheel aqueduct, or you can
go up and round the top of the tunnel entrance to inspect all
the features.

The signpost at the gap points the way for 'Roman Fort and
Woodland Paths'. There are many paths and I just give one
circuit to take in Rough Castle and return. This large community woodland is on the restored site of opencast mining.
Allow an hour for this circuit. Keep on at the first cross-paths
which leads up to near the railway line and joins a good path
coming from the east. Turn west along this until a signpost is
reached (just before the path swings down and round left to
go under the railway to a car park). Take the small path right
as indicated for 'Antonine Wall and Roman Fort'.

In a few minutes this leads through to a stile/gate into
the golf-course look of the Roman site. The path heads left
under the obvious ramparts of the fort to lead to the western
entrance with its subtle defenses. Enter! From there and over
left you obtain good views of the *vallum* as it runs down
and up the hollow of the Rowan Tree Burn. Continue to
the north entrance (another good interpretive board) and

The lilia *(booby trap) at the Rough Castle, Roman Fort.*

from it head out leftwards to another similar interpretive board which pinpoints a special feature of Rough Castle, a strange area of closely-packed pits. This is the *lilia*, dug as a booby trap against mounted raiders. With sharpened stakes set in the bottom of the pits and their presence disguised, the pits would be a very effective way of breaking a charge. There is nothing much visible in the fort itself except vague shapes in the turf. The site was excavated in 1909, but then the foundations were covered over again to preserve them. Objects found during this excavation are in the Royal Museum of Scotland, Edinburgh, where they are, with other Roman remains, given a room to themselves and make a fascinating display.

To return to the Wheel, walk east along the north side of the ditch to reach a stile for the woodland paths beyond. At a fork keep on ('Woodland Walks' are indicated on the signpost) to come to a dip and rise to cross a burn (ignore a bikers' path up an old tip, right). Keep on at a rather confused area—up the bank to find the path clearer again. (The *vallum* is never far away, so the line can't be lost!) Suddenly you pop out onto the path from the starting gap—which can be glimpsed. This circuit can also be made starting from the Wheel.

Another circular walk can be made by continuing west-wards from Rough Castle, a clear ditch section, to Bonnyside House, which also sits on top of the Roman line where the Historic Scotland area ends. Track and minor road lead in to Bonnybridge and the canal can be walked back to the Falkirk Wheel.

In the days of cattle droving, Rough Castle was for one period the site of the Falkirk Tryst, which had started near Polmont after overtaking Crieff as the main market. About 1785, possibly because the new Forth & Clyde Canal made access difficult, it finally moved to Stenhousemuir. The coming of the railways and changes in agricultural methods saw droving largely die out in the 1860s.

Seabegs Wood

This, for canal walkers, is best reached by a curious pedestrian underpass a kilometre or so west of Bonnybridge. You may have to duck your head to get through. Car drivers can reach it from Bonnybridge (east) or Castlecary (west) along the B816, but there is no parking place, and the towpath approach is the safest.

Seabegs Wood, an oakwood, has been well tidied and gives another good idea of the frontier line. The military road has its best-preserved length here, the wall line is obvious and the upthrown ditch material makes quite a rampart. There is an interpretive board (Historic Scotland); altogether a pleasant spot. But how history can turn on little accidents: Robert the Bruce was nearly killed when hunting here in 1300 when he was attacked by a wild white bull.

11
FALKIRK WHEEL TO AUCHINSTARRY

OSLR 65, 64; OSE 349

There's plenty of staging as we leave the Falkirk Wheel site. A double railway line is crossed and a track down leads to parking and 'park and ride' facilities. The railway came 70 years after the canal, so burrowing below the canal was a major undertaking. We are soon back to rural walking, with birchwood across the water on what was mining dereliction, and fields to the right. Pylon lines cross to a substation to the north; just before the first a burn flows under which has the yellowy colour of old mine workings, by the second there is an overspill. Most people are struck by the size of the canal here if they have only been on the Union Canal. The Forth & Clyde appears huge, but it was designed to take sea-going vessels. After an open stretch there are fine trees on the left as the outskirts of Bonnybridge are reached. Several of the first houses have created a varied and attractive shrub garden beside the towpath.

We swing round Cowden Hill (the trig. only 59 metres, but a good viewpoint) to reach one of the canal's lifting bridges (hence the traffic lights) which was opened in 2002. By the gate nearing the bridge there is one of the old oval mile markers: 'Underwood Lock, 2 miles, Falkirk 5½ miles'. We'll see more, though, just to confuse, modern markers may be metric.

Bonnybridge has shops just down the brae, turning left, but another feature that few books even mention, is a must for seeing. Head down the brae and turn right at the Mill

Garage. You will see a small tunnel with another yellowy stream flowing out of it. There is a raised walkway through this pend which gave workers access before the road bridge was built over the canal. It is worth going through after reading the note on the arch: 'The Radical Pend. Named to commemorate the Battle of Bonnymuir. April 5th 1820'. Depression years and hard working conditions in the post-Napoleonic years gave rise to demands for universal suffrage and parliamentary reform etc. Demonstrations were suppressed violently, as in the 1819 Peterloo Massacre in Manchester. Here, a group of 300 Calton (Glasgow) weavers got caught up in a march to Falkirk and were met by mounted troops, who soon dispersed the crowd. Battle it was not, but three were executed for sedition and others transported. Hard to envisage Bonnybridge as a sprawl of industrial works, the last of which only vanished in my lifetime.

Once through the pend, turn right to walk round the brick houses and so back to the canal. Continuing, Bonnybridge sprawls on, on the south side, while below, north, flows the Bonny Water. This rises south-east of Cumbernauld, bisects that town as the Vault Glen and passes under the canal just west of the A80 before continuing parallel to the canal (to the north) and finally joining the River Carron north-east of Bonnybridge.

Somewhere along here Royal Marine Engineers were engaged in laying an Admiralty pipeline in 1918 when the extraordinary picture of their floating barracks was taken (see opposite).

A footbridge is seen crossing the Bonny and the path comes up towards the canal. Keep an eye on it: the path runs along below the towpath and then seemingly vanishes. There is another pend (tunnel) going under the canal, another established right-of-way for miners going to work. It conveniently comes out beside the display board at the entrance to Seabegs Wood—which can be seen from the towpath. Seabegs Wood is another prime Roman site, so do go under the canal to visit. A path of slabs

A unique picture of First World War soldiers with their barrack boats. (© Guthrie Hutton)

keeps the feet dry in the watery pend. The grassy site is studded with oak trees, wall and ditch line obvious, the upthrown *vallum* material forming an obvious rampart. The Military Way can also be made out.

Back on the towpath, another kilometre brings us to Lock 17, where the stable building has been turned into a restaurant and 'Bar 17'. Indian cuisine is available. A rather bare section of canal leads on to the two Castlecary Locks (18 and 19), the former with a lock keeper's cottage. The thundering A80/M80 lies ahead, and before it there's an MM-style bridge which leads a very minor road down and across for the benefit of one house! (At one time it was a swing bridge and a more important road.) The bascule bridge motif is on the parapet and a large MM dating symbol marks the brutally practical A80/M80 Castlecary Bridge. Technically the M80 starts just to the north and this is the A80, but one's as nasty as the other. The A80 barged across to block the canal in 1963, the effective beginning of the canal's end, it felt.

Castlecary occupies one of those places where geography influences history: Roman road and wall, canal, railway and modern roads culminating in the M80/A80 all criss-cross in a bewildering array at Castlecary. The auxiliary Roman fort was one of the few built of stone, and excavations yielded many coins, weapons, urns and other items, and also an altar ingratiatingly dedicated to the god Mercury by the Sixth Legion. Castle, roads and railway have largely demolished the site. The sturdy keep that gave the village its name was the seat of the Baillie family, descendants of the Baliols. The Jacobites burned the castle in 1715, but it has been restored as a private house.

Arrowhead, *Sagittaria sagittifolia*, is a plant thriving in the canal here, and in several places back over the last few miles, yet it is not supposed to grow north of the Tyne. The plant will only grow in unpolluted waters, which is something to commend the canal.

Not long past the A80 there are railings (on parapet foundations) which mark where the Red Burn passes under the canal, shortly to be joined by the Bonny Water, which has all the time been wending east as we've been heading west. A path links the canal to Banknock, and a second one links a little further on, thus making a village circular walk. A distillery once lay beside the canal as we come to Wynd-ford (Banknock) Lock 20, which is important as the eastern

Restoration work on Wyndford Locks

lock of the summit pound of the Forth & Clyde Canal. The western end of this summit level is in Glasgow! The Bonny is still below, on the right, and rises in the hills to the north, as do the waters of the infant Kelvin, whose intricacies I've never fathomed, but the watershed between Forth and Clyde lies somewhere along the next reach of canal—an infamous reach historically.

Lock-keeper's cottage and stables (now a house) have survived at Wyndford Lock (pronounced Wined-ford). Walking on there's a stop lock and a spillway. The next small feature is a basin cut back on the south side (why, I've not discovered) but the major, unique, feature is the greater width of the canal for the next couple of miles. This is the stretch of the notorious Dullatur Bog, which the confident builders decided to cross in a straight line. The bog had to be drained as much as could be, water which now becomes the infant River Kelvin, then a huge embankment created. This was sunk 50m/164ft into the bog before it stabilised! Then they could build the canal: the north side and towpath carefully and the water generally finding its own level on the south side, hence the greater, and variable, width. A path runs up to Kelvinhead and Banton from a former jetty. One sensational event marked the work in the Dullatur Bog: a mass exodus of frogs, tens of thousands of them, which swarmed over the countryside to the dismay of everyone. The local minister was not slow to produce sermons drawing Biblical comparisons.

Wildlife treats the whole canal as an elongated loch, but one with a shoreline out of all proportion to the area of water. Perhaps elongated river would be more accurate, for the canal 'flows'; it is not static water. The flow is controlled so that spates don't tear away riverside vegetation, or droughts reduce the banks to smelly disaster. The canal is a unique environment, and its wildlife features will become comfortably familiar to towpath wanderers. Top of the predation chain would appear to be man, as usual, if you can judge by the number of fishermen seen. Pike, another voracious predator, may be seen lurking by the edge of the reeds. They will take ducklings as well as smaller fish, even other pike. Fish weighing 20lb/9kg have been caught. Perch and roach are also long established, and tench and bream have been introduced. In the Dullatur Moss, a dead trooper was reputedly found still sitting in the saddle of the horse on which he'd fled from the Battle of Kilsyth. The bog had engulfed them. Baillie, the Covenanting general, nearly came to grief in the bog too, but struggled through to Castle Cary, then owned by a cousin.

Witchcraft and superstition lingered long in this area; into the nineteenth century there were known 'witches'. A sceptical farmer met one when carting along the canal bank and gave her a piece of his mind. She held up her fingers and muttered a curse before taking to the fields. The farmer laughed and plodded on. The sedate mare, however, suddenly went daft and plunged, cart, farmer and all, into the canal. (One suspects the farmer had to have some excuse.)

The most interesting hill sections of Roman wall lie above the canal from Craigmarloch westwards, and give fine views (of the canal amongst the rest) so deserve a walk too, with a bit more effort than along flat towpath! They are described in the next chapter, then Chapter 13 will explore Kilsyth. Chapter 14 takes the canal into Glasgow.

From Craigmarloch the canal runs along under wooded Croy Hill with many a wiggle, then there is a pleasant stretch on to Auchinstarry with long staging leading to a big basin, which is a major marina with all the facilities, as well as being a Lowland Canals depot. The Auchinstarry Bridge is another built to slope upwards, replacing a swing bridge, the abutments of which lie west of the new bridge, on the old road line.

The Antonine Walkway Trust have created a range of paths over the hills and linking with the canal. On our side the eye is drawn to a large whinstone quarry, the cliffs surrounding a deep pond, used for fishing instruction, as the cliffs are for climbing. Picnic tables allow grandstand observation. I once saw the advert for a lecture by a famous climber, entitled 'World Rock', which would cover 'Australia, Yosemite, Italy, France and Auchinstarry Quarry'.

Auchinstarry was where Kilsyth coal would be loaded onto canal boats to take it to Glasgow and even to Belfast. Later, the mineral lines all converged on Twechar, and then railways completely took over from the canal. Auchinstarry quarry produced the whinstone 'setts'

A climber in action at Auchinstarry.

that paved the streets of Glasgow, and dates back to the 18th century. Kilsyth lies north of Auchinstarry Bridge; to the west is the wetland bird reserve of Dumbreck Marsh.

A pylon line joins us at the Kelvinhead jetty and crosses just before the next bridge, another with historical associations. The long straight does end! From Victorian times to the demise of canal cruises, Craigmarloch was a popular destination from Port Dundas in Glasgow. Old photographs show 'The Bungalow' (restaurant/tearoom, putting green, swings, etc) with several boats, the various 'Queens', (*Fairy Queen, May Queen, Gypsy Queen*), in the basin. All buildings and signs of this past activity have gone, and the basin, filled with reeds, is barely discernible on the south side just before the bridge. The old bascule has gone, and the modern one is just functional concrete, built on an upward angle.

There's still plenty of interest here as the display boards indicate. Now almost overgrown with trees, the ruin well to the north of the bridge is a former stable block. In winter it can be glimpsed looking up the canal feeder. The site is so far off because of the boggy ground; an earlier stable building just sank into the mire! A quarry behind the stable supplied stone for the canal. The feeder burn comes down from Banton (Townhead) Loch, a reservoir, which explains the 1645 battle symbol in the water. (More about the battle later, but names like Baggage Knowe and Slaughter Howe also pinpoint it on the map.) The feeder circles the hill, passes the stable ruins, and enters here. Just west of the bridge is an overflow from the canal, the canal being higher than the River Kelvin and the large alluvial plain stretching over towards Kilsyth. This feeder is the main water source for the Forth & Clyde Canal. (A lock spills 80,000 gallons of water every time it is used, so topping-up is essential.) The reservoir taps the Birkenburn Reservoir and Garrel Burn coming off the hills above Kilsyth. An attractive line of cottages west of the bridge was short-sightedly demolished in 1976.

Gypsy Queen *at Craigmarloch*

A busy scene at Craigmarloch in the years of popular tourism.
(©East Dumbartonshire Council, W.M. Patrick Library
Kirkintilloch)

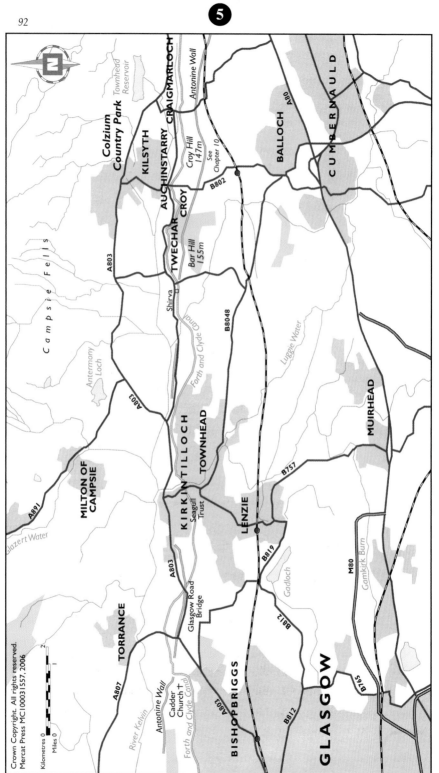

Kilometres 0   1   2

Miles 0   1

12
THE ROMAN WALL: CROY HILL AND BAR HILL

OSLR 64; OSE 349

With this hilly ground available the Romans, unsurprisingly, ran their defence line along the crest of the hills. Until comparatively recent times low ground across Scotland's central belt was boggy in many places, one reason the canal was such a boon to travel: compare its ease to a coach 'claistering through the glaur' as was the usual inter-city travellers' complaint. Croy Hill and Bar Hill can be done separately or together as pleasant circular walks in conjunction with the canal.

Start at Craigmarloch Bridge and walk up the Dullatur/Cumbernauld road. About 100m before Wester Dullatur Farm turn off right, where there is a Historic Scotland sign for 'Croy Hill ¼ mile'. The track leads to a gate and then on past a big pylon. Turn left at a junction (picnic site, right). The line of the Roman wall/ditch can be seen crossing in front of the next pylon, and you follow the track round to pick up its line. (Look for the Historic Scotland notice.) The railway will probably be heard if not seen—your last contact with this inter-city artery which passes south of Croy Hill in a deep, mile-long cutting. An alternative line, the Kelvin Valley Railway, wound along by the hills to the north, linked with the Kilsyth and Bonnybridge line at Kilsyth. (John Thomas's *Forgotten Railways,* has interesting stories on Scotland.) Coal mining, on Croy Hill and Bar Hill, at Twechar, Shirva, St Flannan, Tintock, Cadder, all on or near the canal, has vanished with almost no trace. There's still an active quarry on the south flank of Croy Hill.

A path winds on, up towards a clump of trees which marks the site of a Roman fort, or you can follow the deep V of the ditch. Just before the sycamore clump there is no ditch, the rock being too hard even for the Roman soldiers to quarry and, later, the scarp itself is an adequate defensive feature. On the east summit of Croy Hill there is a big view, as you would expect, but most eye-catching is the ditch which arcs across the side of the marginally higher western top. From there, descend a spur (the line of the Roman wall) towards the north edge of Croy. On the descent to Croy the ditch is actually hewn out of solid rock (it shows clearly, looking back from Bar Hill). There are good views down to the crowded Auchenstarry Basin.

Croy is a rather sad town, for its mines closed in the early 1980s and nothing has taken their place. The path descends to skirt the north end of the town, passing the sports hall of the Croy Miners' Welfare Charitable Society and joining the B802 at Croy Tavern. A short distance along the main street on the left is a small snackbar which may be welcome but it shuts by mid-afternoon.

The B802 is dangerously busy and the brief walk down it to the canal can be avoided. Before reaching the Croy Tavern you would have seen a signpost indicating whence we've come and the way on for Bar Hill and Twechar. Forty metres on northwards turn left by a transformer through a green kissing gate for the continuation but, if intending to descend to Auchinstarry and/or Kilsyth, don't turn off but continue on a safe cross-country path that wends down to reach the canal which is followed along to the basin/marina. The towpath, of course, is on the north side from there.

Keep to the field edge from the kissing gate to reach another onto the B802. A sign for Bar Hill indicates the track across the road. Take this farm track (which runs up along the Roman line), ignoring left forks, on to a gate leading into a cool deciduous wood. Ignore paths leading off to the left. When an open green space is reached (the green swathe, straight on up the forest ride, is the

The sweep of the Roman ditch on Castle Hill

line of the Military Way), bear right up to a bump where there is a sign for Bar Hill Fort and an explanation of the Roman period.

The route is now a wide green sweep; forest on the left and the deep V of Roman ditch below. The ground dips and climbs steeply and the easiest walking is by the forest edge. At the top end of the wood a final pull up a cone of hill lands the walker at the prominent trig point of Castle Hill (155m/508ft), the highest part of the whole defensive system and, naturally, a superb viewpoint. On a clear day both Forth and Clyde waters can be seen, and the hills to the north are spread in fine array beyond the ever-expanding sprawl of Kilsyth. Castle Hill's summit has traces of an Iron Age fort and, continuing west, there's a Historic Scotland interpretation board.

Wend on through open woodland, left rather than straight ahead, to arrive on top of Bar Hill where the outline of the Roman fort has been preserved. The setting is splendid with the view to the Campsies running westwards to the dip of the Blane and then on to the Kilpatrick Hills, and with the spread of Cumbernauld and Glasgow a contrast to the south. The site is well explained. When the Romans departed all manner of things were thrown down to choke the well (a treasure trove for archaeologists a millennium later), including the winding gear of the well itself, an altar, weapons, tools, ballista balls, over 20 columns, bases and capitals from the headquarter's building, pottery and much else, all preserved till dug out in the twentieth century. (Some finds are in the museum at Kirkintilloch, others in Edinburgh.) Bar Hill Fort is unusual in being sited back from the wall, so the Military Way passes between fort and wall/ditch.

Another notice-board, down to the north-west, explains the ruins of the separate bath-house and latrines. Every fort had its bath-house, laid out in roughly similar fashion. You entered a changing room, then a cold wash room before a series of ever-hotter bath rooms. The heat came from under-floor and/or inter-wall ducts, the air being heated in a furnace at the end of the building. The toilet block would usually be sited downhill slightly so waste water could be channelled down to flush it. A trip to the toilet was a sociable event—the room was fitted with the optimum number of wooden seats ranged round the walls and lined over troughs. All very hygienic and organised, and very Roman. One find here was a rubbish pit with things like broken pottery, old boots and eleven human hand and foot bones. Surgeon's work or what? I'm still puzzled as to how they provided enough water for the site.

Continue from the fort, descending towards the south west to a gate with the round dome of a water tank very visible. A second gate exits onto a farm track. Turn right to follow this down to Twechar, coming to the main road at the war memorial and the Enterprise Park opposite. A short distance along the road, left, is the minimalist Barrhill Tavern. There is little else, apart from a small shop. Descend towards the canal then. The houses passed are those of the staff of the coal mines, the teacher, minister and such. The miners' rows have long been demolished and basically a new town of council houses created further west. Glenshirva Road beside the canal leads to the shop and the canal description is resumed at Auchinstarry; in Chapter 14.

Opposite: the Roman Bath House remains on Bar Hill, Twechar and the Canal in the distance.

13
HISTORIC KILSYTH

OSLR 64; OSE 349 or 348

Following the B802 from Auchinstarry into Kilsyth, you see
the town sign bearing the burgh's coat-of-arms, and, just
beyond, is an unusual watch-house in the cemetery on the
left. Under this watch-house lies the Kilsyth family vault,
of which a gruesome story will be told later. On entering
the graveyard, note the unusual lamb sculptures on stones
to left and right, memorials to young girls and a change
from the ubiquitous draped urn, which is even present on
a cast iron 'stone'. There are also several interesting older
stones (some lying flat), obviously the work of one mason,
filled with symbols: trumpet, open book, hourglass, shut-
tle, a ship under sail and topped by a winged spirit with
angular wings.

The road into Kilsyth dips to playing fields, then pulls up
again. Have a look at the monument on the right, erected to
the memory of a minister who died in 1910. Apparently Mr
Jeffrey 'wore the White Flower of a blameless life'. Kilsyth has
a tradition of religious fervour dating back to Covenanting
times, which became most notable during religious revivals
in the mid-18th and 19th centuries. Few towns have as many
churches. Pass the decoratively planted roundabout and then
turn down right on a pedestrian path which leads into the
heart of the town.

Kilsyth has become a vast sprawl of modern houses, but
still maintains a heart of historical interest. The library on
Burngreen (open during office hours) usually has historical/
local displays, and the green itself has a painted lady fountain
(akin to the one in Tomintoul, or Kirkintilloch's Peel Park),

a bandstand, bridge railings (each different) and even house signs (one a tortoise), all made of cast iron. Just off the green is the attractive Market Square and a Main/ High Street which is being brought back to life.

Kilsyth has gone through hard times, even recently, when the mining industry finally petered out. The library has a display on this more recent history. An exhibition on the local temperance movement had a cutting from the local paper

A commemorative memorial in Kilsyth

in 1923 (when the town went 'dry') pointing out that the sale of methylated spirits had increased by leaps and bounds!

East of Kilsyth is Colzium Park and its mansion, which was given to the town in 1937. In Colzium House there's a display on the Battle of Kilsyth. A plinth below the house commemorates the battle, but boobs in ascribing the victory to the Duke (sic) of Montrose. (The Grahams only gained that title through supporting the 1707 Union sell-out, and the local Livingston, Lord Kilsyth, fought its every clause.) The grounds are well-kept and full of interesting shrubs and trees, glorious in autumn colours. The immaculate walled garden is a gem. The old laundry has been restored and turned into the Clock Theatre, named after the clock above it which dates from 1863. Just up from the bridge, leading to the house, is a well-preserved example of an ice house. There's the oddity too of an animal cemetery(now signed 'Full'). Townhead (Banton) Reservoir lies just below and to the east, and is the main 'feeder' for the Forth & Clyde Canal.

In 1739, the Colzium estate factor was responsible for introducing the potato to Scotland. Robert Graham began experimenting with potatoes in his garden near Banton and planted out crops above Kilsyth. An astute business man, he bought up farms right across Scotland and planted potatoes. As they say, the rest is history.

To revert to earlier history, Kilsyth 1645 was the last of Montrose's victories in his *annus mirabilis*. While Montrose was in north-east Scotland, Cromwell had inflicted a crushing defeat on the King's forces at Naseby, and Montrose realised that however many Covenanting armies he defeated in the north, the result would be marginal. He had to move south. In the middle of August the two opposing forces found themselves face to face at Kilsyth.

The Covenanters' general, Baillie, had the larger force, but was hampered by his serving a committee which included the Earls of Argyll, Elcho, Burleigh and Balcarres, who had all suffered under Montrose's Highlanders. Montrose was encamped below the Campsies, the Covenanters were on a ridge above. Quite why Montrose had allowed his opponents the higher ground is not known, but the ground between was not suitable for cavalry and Baillie was wary about a possible trap. His committee, however, felt they had caught Montrose at last, and to leave him no chance of escape they began to shift their men across Montrose's front to occupy a dominant hill from which they would swoop down.

In a glen below the route, a small force of Macleans occupied some houses, and a small body from the Covenanters' column broke off to attack them, were repulsed and chased back. This was too much for Colkittos's men, who charged after them. The clansmen swept through the column and before long, the Covenanting army was in flight. Scotland belonged to Montrose, the King's Captain General. But south of the border there was no Montrose, and his defeat at Philiphaugh near Selkirk was only a month away. He was to lose his wife and eldest son the same year. He was only 38 when he was executed. Five years after the Battle of Kilsyth, Cromwell marched into Scotland and, following his 'crowning mercy' of victory at the Battle of Dunbar, carried on into the west. On the way he blew up the Livingston castle at Kilsyth.

The Livingston support for the Stewarts was to have a somewhat macabre continuation. The ill-fated Bonnie Dundee, who died in his moment of victory at the Battle of Killiecrankie (1689), had married Jean Cochrane, grand-daughter of the Earl of Dundonald, who then, as his widow, married William Livingston (later Viscount Kilsyth) with whom she went into exile. On a visit to Rotterdam in 1695 the couple were making a goodnight visit to their infant, asleep with her nurse, when the roof fell in and all but Livingston were killed. The bodies were embalmed and brought home to Kilsyth. In 1795 the vault was accidentally opened by some students who had found a way in and the bodies were found to be remarkably well preserved. They became something of a spectacle before the vault was closed again. I found this in a book dated 1872 and visited the site in 1989, a year when another, official, inspection showed the bodies were still preserved—three hundred years on.

The *Statistical Account* is very thorough on Kilsyth, and full of fascinating descriptions of work generally, and particular topics like the creation of a cut to take the River Kelvin, or the coming of the canal. The minister describes the local climate as 'rather watery'. In 1733 the area was hit by a freak thunderstorm when three-inch hailstones wrought havoc and left the country under water. The burns came down off the hills in torrents and did a great deal of damage (20-ton boulders were trundled down) but, amazingly, no human life was lost. A woman and child had a fright when a bolt of lightning came down the chimney as they sat close to the fire and killed the unfortunate cat at their feet.

6

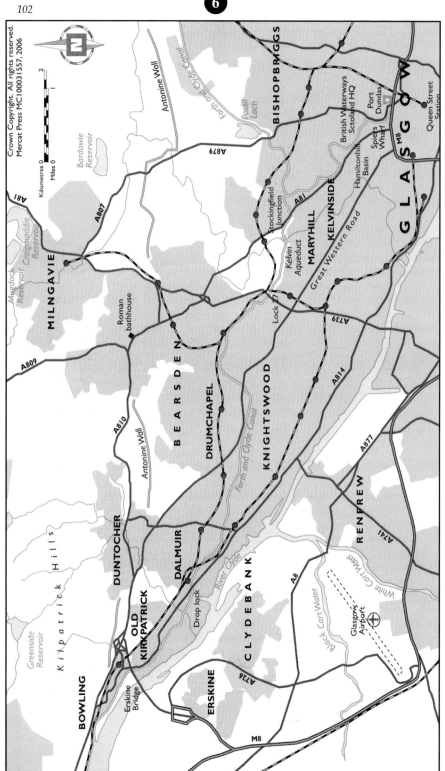

14
AUCHINSTARRY TO KIRKINTILLOCH

OSLR 64; OSE 348, 342

A fine curve of stonework sets us off from the site of the old swing bridge and the attendant cottage for an even twistier wending below Castle Hill/Bar Hill. The canal is strongly walled in places as it holds the heights. All signs of the once extensive mines and quarries have disappeared in the spread of trees. Yellow flags and blue damselflies brighten summer reflections, a rich opulence. However we are soon walking along above the B8023, Kilsyth still a sprawl visible below the Campsies and Queenzieburn west of it. Eventually the towpath and road are merely separated by a railing, and for the last 70 metres to the Twechar Bridge there is a thin strip of pavement. Ca' canny! The bridge is a lifting bridge and alongside it are the abutments of an earlier swing bridge.

Twechar is almost unpronounceable unless born there; try twek-char (the 1823 *Companion* had Quechar). It never starts by rhyming with wheech. Created as a mining village in 1860, its original miners' rows have gone and the modern council houses sprawl westwards. The school and church and better old buildings lie up the road south, where walkers descending from Bar Hill with its Roman fort come out. The map is littered with 'Quarry (dis)' indications. One pit, just east of Twechar Bridge (north side), closed in 1964 and the last in the area in 1968. The canal had at one period been bridged west of Twechar Bridge to link collieries on the condition coal was moved by barges.

There are two options for continuing: to keep along the towpath for a stretch still tight against the B8023, until the road is forced away by Shirva Farm. The towpath is backed by gabions as it swings left to pass the farm; the alternative, recommended, is to walk the south side of the canal as far as Shirva Farm. (The canal here lies on the Roman Wall line.) Walk along the road parallel to the canal, passing a village store, and when the tarred road swings left into the housing scheme, keep ahead on a track which runs along the backs of houses, with many garages. At a fork bear right. Soon there is another fork with a footpath, right, obviously going onto the canal bank. Do not take this, but continue down the sunken lane (the *vallum*) to its end at a stream, the Broad Burn, where there is a footbridge. There are also several paths, but ignore these and turn downstream to go through the tunnel under the canal. There is a date, 1771, above the arch at the north end. At Shirva Farm and beyond there is not much sign of the Roman line, which then vanishes into Kirkintilloch. (There is a south bank path all the way through to Kirkintilloch from Twechar.)

The pend has a raised walkway and a larger stream than the Radical Pend. Its practical use has long gone. Worn stone steps take us up to the towpath. If coming along on the towpath, the red painted railings will be noted just past the farm where the ground falls away steeply.

Continuing, there is another canal stable block, sadly no more than a dangerous ruin. A minor overspill follows. More puzzling is what looks like a Union Canal milestone with the letters FCN inscribed. (Forth & Clyde Navigation; probably a boundary stone originally.)

The hamlet of Tintock lies across the canal, once a scattered weavers' base in a lonely setting, later with several pits nearby, one with a mineral line to the canal. There is a staging by the old cottage and modern bungalow, and possibly a boat may be moored, recalling days of the 'Queens'. The Kirkintilloch suburb of Harestanes now spreads out to Tintock. When opposite a group of three rectangular blocks of flats, look out for a path sloping back down from the towpath. This leads to the Tintock pend arch with its low headroom of 4'9" (1.3m), still used by cars.

At a bend a path angles down to the junction of the B8023 with the A803 Kilsyth road which crosses the River Kelvin. The river flows on to skirt the north of Kirkintilloch, adding the Glazert Water which drains the western Campsies and the Luggie Water flowing from Cumbernauld. As a contorted big flow on a flood plain, it gives marvellous examples of oxbow lakes. The Roman Wall too has crossed to run along the south side of the canal. A pipe is bridged over the canal. Houses now lie down to the right, dominated by old trees. On the south side is the bold brick St Flannan's Roman Catholic church, well worth a visit as it's only five minutes' walk from Hillhead Bridge, and the airy, dramatic, modern interior is beautiful.

We finally come on the 1938 Hillhead swing bridge with its ironwork and lantern finials. The church tower of St Mary's and the steeple of St David's have become landmarks ahead. The canal's greater width goes back to Kirki's days as Scotland's first inland port (1773). Herring might be sold off boats which came through from the Forth (5p a pailful),

An ice breaker helping traffic on the canal under the Campsie Fells.

there were timber yards, and a thriving import/export business, besides local foundries and mines. The town would have had mixed feelings when the canal was cut westwards to Glasgow.

When fine railings draw attention there is a notable engineering feature, the huge single arch over the Luggie Water, but note how a railway goes under the arch as well. This Campsie Branch Railway was added in the 1840s, which hardly detracts from the massive yet graceful arch—well worth going down to see. A decorative pillar stands by the towpath.

A concrete flyover (Nicholson Bridge) follows, stark on its round pillars, noisy with 'doos' and the canal busy with ducks and swans who receive hand-outs galore. St Mary's (red stone) parish church was built in 1914 to replace the Auld Kirk, which is now the excellent museum. There's another decorative pillar with the Cowgate (the town's main street) crossing the Townhead Bridge. The original bascule bridge had been replaced in the 1930s with a swing bridge, and then the canal was culverted in the destructive Sixties. The present plain concrete bridge rests on the old abutments. The building over on the corner of the Cowgate was the Eagle Inn during the 19th century, where the owner, Sandy Taylor, supplied horses for towing and operated passenger boats. Kirkintilloch is more or less the middle of the long summit pound. In 1836, about quarter of a million passengers were carried city to city, commercial craft passed constantly, even the herring fleet might hurry through from Clyde to Forth after the 'silver darlings'. Children living in Kirki had a great playground. By turning right down the Cowgate

Kirkintilloch motto on a fountain.

you can find most of the town's services. The next chapter takes the canal on from Kirki right into the heart of Glasgow.

Kirkintillock's name, anciently, was Caerpentaloch, the fort at the head of the ridge. Note the well with the motto over it: 'Ca' canny but ca' awa'. The centre is modern and has little character, but walk to the Cross at the far end to visit the Auld Kirk (1644) Museum containing award-winning displays on the town's major past industries: coalmining, iron works, shipbuilding and weaving. The Lion Foundry made the red telephone boxes which were known all over the world, now sadly phased out and the works closed.

The Barony Chambers next door dates to 1815 when it replaced the old tolbooth. The top floor had a school, the middle floor acted as town hall and court room and below lay the gaol. The steeple's clock was known as the 'four faced liar', as each face tended to show a different time. The museum has an interesting range of local books etc., for sale.

Behind the Auld Kirk are commemorative gates leading into Peel Park, where there is a fountain and bandstand (as at Kilsyth), a good view to the Campsies, an excavated section of Wall foundation and the site of a castle 'motte'. Walk down the park to Union Street and turn left back to the Cowgate. There's another red church (St Ninian's Roman Catholic) beside Peel Park. Kirkintilloch, like Kilsyth, has plenty of churches, sharing the same history of religious revivals in the 18th and 19th centuries. South of the canal lies Townhead; the only interest there is a couple of pubs/restaurants.

Kirkintilloch developed as a result of the canal coming. Two shipyards were sited here and industries developed as they could use the canal for transport. Glasgow traded with eastern Europe via the canal. Maryhill (Kelvinlock then) had the first registered Temperance Society in 1827, and Kirkintilloch was also a dry town for 47 years, between 1921 and 1968. Records still show that many of the accidents on the canal had alcohol to blame—drunk in charge of a scow, screw or gabbart perhaps! Something like three million tons of goods and 200,000 passengers were being carried annually in mid-Victorian times.

From late Victorian times until World War II, cruising on the canal was popular (an alternative to sailing 'doon the watter'). The famous '*Queens*' regularly ran to the basin at Craigmarloch (Kilsyth) with its tearoom and putting green. A band and singing enlivened the evening run home to Glasgow. This is the image which lingers, rather than the reality of the canals being, for 200 years, as vital to industrial transportation as are the motorways today.

Robert the Bruce gave Kirkintilloch's castle and lands to the Flemings, but they rebuilt the castle at Cumbernauld as base. Edward I seized the castle, but Bishop Wishart paused in his building work on Glasgow Cathedral to help dislodge the English forces. Bonnie Prince Charlie's army passed through Kirkintilloch on the way to Falkirk. A shot was fired after the troops, who had to be placated to stop them sacking the town. The library has a treasured burgh 'Court Book' (1658-94) which contains a range of historical information. In those days it was a crime in Kirki to be unemployed.

Kirkintilloch was early involved in railways as well as canals, a line being built from the mines at Monkland in 1826 (for horse-drawn wagons) to take coal by the canal to Edinburgh. Lines proliferated, and in 1840 another went by Slamannan to the Union Canal at Causwayend. North of Kirkintilloch the line skirted the Campsies, so you could travel by train to Aberfoyle, Balloch and Stirling—all lines which have gone now.

This book has had to use an odd compromise between metric and imperial measures, thanks to the mess that we made of metrification, but change is not new. While browsing through the *Statistical Account* on Kirkintilloch, I saw that the minister quoted measures (and money) which now mean nothing to us. 'A chalder of lime, consisting of 16 bolls, each of which contains 3 firlots is bought for 6s 8d'.

Kirkintilloch had one remarkable visitor in 1785: the Italian balloonist Vicenzo Lunardi took off from central Glasgow watched by a crowd estimated at 100,000. He landed north of Kirkintilloch, which practically emptied as people streamed out to see this marvel. A mile or two from Lunardi's touch-down site is Antermony where a more remarkable traveller was born in 1691. John Bell, a doctor, went to serve the Czar Peter I in 1714. From there he went to Persia (Iran) and later right across Russia to China—a 16 month journey. After further journeys in Turkey and Persia he became a merchant in Constantinople. He returned to Antermony about 1746, and wrote a book about his adventures. An unusual canal visitor in 1952 was the midget submarine XE IX, which spent the night in the J & J Hay boatyard.

Probably the town's most famous son was Tom Johnston —journalist, historian, politician, Secretary of State for Scotland, creator of the Highland Hydro industry, chairman of the Scottish Council of the Forestry Commission and the Scottish Tourist Board (among other things). He was also a consummate politician and a member of Churchill's wartime cabinet. It was the war that led to his appointment as surely the best (some would say only) Secretary of State *for* Scotland, and out of the war, with shortages and difficulties, the Hydro Board was created to tap the Highland water for power, a fairly assured renewable energy resource.

Kirkintilloch is another place where water is fed into the canal to keep it operational, this time coming from the Johnstone, Woodend, Lochend and Bishops Lochs, through which the northern end of the M73 passes. The canal was made waterproof by lining it with a coating of 'puddled' clay, though the graceful Union Canal aqueducts are simply iron troughs.

15
FROM KIRKINTILLOCH INTO GLASGOW

OSLR 64; OSE 342

We continue with Glasgow 11 km (7mls) ahead and Falkirk 25km (15½mls) astern. Heading west from Townhead Bridge, it is difficult to envisage this area with a busy shipbuilding industry, but something like 150 boats (scows, puffers and the like) were built at two yards. Puffers were launched sideways with a great splash, a spectacle that always drew a crowd. One yard was sited where you still see a large boathouse, the other in a huge basin further back (now filled in). The present boathouse is the quarters for the Seagull Trust boats *Yarrow Seagull* and *Marjorie Seagull*. The books of A. I. Bowman (see bibliography) give more fascinating details, and look at the G. Hutton historical picture titles too. The last launching was made in 1945. There is also much on display at the museum. And some dramatic changes are in the offing. The large building on the south bank with all the railings is the Kirkintilloch Learning Centre, and just west

A boat launch in 1895, Kirkintilloch

of it is a big launching slip, not easily seen from the towpath side, where the original filled-in basin is to be recreated as a modern marina. As early as 1826 a railway was laid from Monkland to Kirki so coal could be dispatched eastwards efficiently. In hard winters all traffic on the canal would come to a standstill until a thaw materialised. Railways proved superior.

A bend leads round a bluff to pass St Ninian's High School. Across lies Joe's Wharf with attendant boats no doubt, perhaps including *Wee Spark*, the 1:3 replica of the *Vital Spark* of Para Handy fame. A big wall marks the small Park Burn overspill. A firm towpath through a rural landscape gives easy walking to Glasgow Road Bridge. A plaque on the wall below the bridge mentions it being opened in 1990—the first of the culverts to be replaced with a bridge so canal navigation could be resumed. (The original was a swing bridge.) The bridge is concrete, but with attractive lines. The Stables was just that in olden days. Its modern status as a restaurant and pub may be welcome! There are seats outside and extensive car parks. The Forth & Clyde Canal Society has boats based here and there is the 'Craft Daft' studio boat. You may see the *Gypsy Princess*, *Janet Telford* or *Voyager* if they are not off with school parties or Santa Cruises! Moving on, there's a dock and launch slip opposite, though the earliest canal boats had to be lowered by crane!

The Roman Wall line runs parallel to the A803, so the canal cuts it once more just east of Glasgow Road Bridge. The wall crosses back at Cadder, then keeps well north through Bearsden (the foundations of a bathhouse on display) before a gradual descent to the Clyde at Old Kilpatrick. For many centuries it was called Graham's Dyke, or Grim's Dyke, just as Hadrian's Wall was the Picts' Wall. Archaeology is a young science.

The din of the A803 fades as the towpath wends westwards, a remarkably rural section, and the rural feel is maintained for a surprisingly long time, if pylons are ignored. The bridge on the Torrance road, Hungryside Bridge (a 'drawbridge' originally), is rather battered and repaired, but loved of cushie doos. And why the 'windows' in the abutments? There's a small car park. The Campsie Fells and the Kelvin valley fill the view to the north.

The canal twists and bends round to reach Cadder Bridge, another drawbridge originally. (Cadder is pronounced Cawder.) The Church is big and dominant (and busy if you arrive on Sunday morning) and in the grounds there's a mort safe and watch house harking back to the days of the Resurrectionists. With prices for corpses higher in Edinburgh than in Glasgow the canals came in useful. Cawdor Golf Course surrounds the site and Cawdermill House next to the bridge has been restored.

An underground fire in Cadder No 15 pit cost the lives of 22 miners in 1913. Two railway accidents are connected with the Cadder stretch of the Edinburgh-Glasgow line. In 1973, fifty yards of track were ripped up during a high-speed derailment, and ten years later passengers had to leap from the train as two coaches caught fire.

In the 18th century the patronage row in the Church of Scotland involved a young lawyer who was an elder at Cadder Kirk, and whose family home at Huntershill has made the name Muir of Huntershill one to remember. He suggested the idea, to us quite innocuous, that everyone should have a vote, and for his stand on such principles he was to be thrown into a life which, if written as fiction, would sound improbable.

Arrested for sedition and then released on bail, he went to France to try and plead with the Revolution leaders not to execute Louis XVI, as that would damage the cause of reform in Britain. The outbreak of war between Britain and France meant he was late back for his trial, so he was arrested and carried to Edinburgh in chains. In England he would probably soon have been free, but the grim Lord Braxfield sentenced him to 14 years transportation, which was all too often a death sentence in disguise. Muir survived the horrendous voyage out to Australia and then escaped on an American ship

Sunset on the canal near Cadder

which, after crossing the Pacific, was wrecked on the west coast of North America. He and one other survived, though the latter soon died.

Despite hostile Red Indians, Muir walked out of that situation only to be arrested by the Spanish authorities in Central America. He was sent to Spain, but that ship was wrecked too. Fished out of the sea, he was taken to Cuba and then across the Atlantic to Cadiz, where the ship was engaged in a ferocious battle with an English frigate. Muir fought for the Spaniards as the lesser evil, and was severely wounded, losing the sight of one eye. Put ashore, he was eventually freed to travel to France, where he lived out the rest of his short but eventful life. Muir was only 34 when he died.

Patronage simply meant the local laird, not the congregation, appointed the minister, an issue that divided Scotland for a century until the system was abolished. In Kirkintilloch those who objected used to walk over the Campsies to church in Stirling each Sunday—a round trip of 35 miles!

Leaving Cadder, the going feels spacious, with a deep ditch on one side and the canal on the other, but the next bridge, Farm Bridge (Balmuildy Road) is no showpiece and the big 'Leisure Dome' is, after all, on the edge of

the sprawl of Bishopbriggs. If not tempting on the day
for swimming, badminton, squash, aerobics or anything
physical, it does offer refreshments. A rather uninspiring
two miles follow, the views north hidden by the slope of
a one-time infill site, with the Wilderness Woods beyond,
the way ahead an array of pylons all converging on a big
electrical substation, and with the first cluster of tower
blocks of which Glasgow has many.

A good section to cycle perhaps. Working canal boats
often carried a cycle, as a crew member had to assist with
locks and bridges and so could be dispatched ahead. Puffers often lowered a man ashore by derrick to save precious
time.

The SWT Possil Loch Nature Reserve, with its extensive
reed beds, offers a break: a footpath heading back, right,
leads into this, and the sanctuary can be circled to exit just
before the next bridge at Lambhill. I saw one roe deer last
time there, and watched a buzzard with a dead rabbit being
harrassed by magpies.

The A879 Lambhill Bridge (to Milngavie) is no joy either.
An old 'horse barracks' (change house) is abandoned and
graffiti all too evident. A couple of minutes south is a post
office and some shops. The mile on to the Stockingfield
Junction is insalubrious, a mix of industry (alive and dead),
uninspired housing, waste ground and litter galore,
although with a few surprising pockets of greenery. A
railway is crossed and the canal twists and turns to confuse
landmarks, the ornate tower of what was Ruchill Hospital
(looking like a Flemish bell tower but infact just a water
tower) will seem to appear in all directions! The junction
follows a derelict stop lock in unobtrusive fashion.

The Carron Sea Lock is a satisfying 26 miles behind,
Bowling 10 miles ahead, but first there is the 2½ miles to
Spiers Wharf and the new addition of the major, bold,
linking back into the system of Port Dundas. We are still
on the summit pound, just, so there are no locks along the
Glasgow Branch to Spiers Wharf. The towpath hereafter is

on the south side (or west on the Glasgow branch) so we, perforce, drop down, go under the low arch of Lochburn Road and climb up again, and only now is the going blatantly urban, and the more interesting for that. One typically British anomaly here is the mix of distances given sometimes in miles, sometimes in kilometres, the latter presumably linked to EU funding of signs. A century ago a ferry saved this down and up, according to an 1898 OS map.

The Glasgow Branch runs high, so there are views over rooftops and streets to spires and towers. A dismantled railway passes below, then a three-arched overspill is passed and a bend leads to Ruchill Street Bridge which, like the Firhill Bridge ahead, had to be rebuilt in MM style after their years of being culverted. A basin (now filled in) once served the Bryant and May match factory, which has also disappeared. Ruchill Church on the right, built of red sandstone, has an adjoining hall built earlier, designed by Rennie Mackintosh. It now faces a Macdonalds! Across the bridge are one-time rubber works built in period brick, which are listed buildings and slowly being restored for other uses. Many other old buildings have disappeared though, and some attractive flats now look onto the canal. The canal itself acts as a healer.

Tenements are passed on the right, with big new apartment blocks up across the canal. Bilsland Drive aqueduct (1879) sees traffic pass unobtrusively under the canal (unless you peer over the parapet) which then makes a big loop with busy Maryhill Road below on the right. A footbridge connects with modern housing. Firhill Road bridge (Nolly Bridge) rises across the canal to run up to Ruchill Park and, off left from it, the sweep of the Murano Street tenements (1899-1903) are quite spectacular. In the crook of the loop lies Firhill Park, home of Partick Thistle Football Club (the Jags) with the big block of Firhill Court (student accommodation) looking onto the canal. At Queen's Cross, on the other side of the stadium is an 1899 Mackintosh church, his first ecclesiastical work, now the home of the Charles

Rennie Mackintosh Society. The Firhill Basins lying on the outer side of the loop were once busy timber yards. Much of the timber was from Scandinavia and brought through the canal from Grangemouth. Timber basins were used for seasoning the wood in water so, as you can imagine, were notorious for causing fatalities among children, who could not resist playing at 'rafts' and who would swim whenever supervision was absent. The canal drowned the careless. There's a story of one local who became quite a hero and was awarded medals for leaping in to rescue people, until it came out that he was getting a mate to push them into the water in the first place. The Murano name may come from the island near Venice, as glassmaking was also a one-time Firhill industry.

There is plenty of staging here but, sadly, nobody is going to leave a boat unprotected in these parts. Here too, in World War Two, another stop lock was added in case enemy bombing breached the canal (which would have been disastrous for so much of the city below) so the canal was carefully portioned off into safer volumes of water.

The canal swings gradually left. Across, inset into Hamilton Hill, is an abandoned basin which was once a clay quarry, source of material for lining the canals. A boarded-up green-painted pigeon loft is all that points to a once popular activity along the canal. The university is well seen, and all the city spires and towers. A wide section of towpath leads us to the oval-shaped Hamiltonhill Basin, with a cluster of boats moored along the enclosure of British Waterways Scotland HQ. This was the original (1777) terminus of the Glasgow Branch. There's an obvious stop lock and a row of workshops, and working boats for dredging weed and all the junk that gets tipped into the canals. The slip at the office block was the first on the canal where the company built and serviced their boats.

A lane heads down, right, leading to Garscube Road, under the M8 and so into the city centre, a 10-15 minutes walk. Rockvilla Bridge, a bascule bridge, of the style to be met frequently on the Bowling section of the canal, leads across to the BW Scotland site on Applecross Street, the street coming in off Possil Road, which goes under the canal and down to the M8, the last major aqueduct. The Possil Road aqueduct dates to 1880, but the original 1790 Whitworth aqueduct which is 'seduced' is easy to overlook. It comes first, a curved wall over which you peer down at a century and more of litter. It is worthwhile dropping down to street level to look at the massive stonework of the aqueduct, similar to those of Bilsland, Lochburn and Maryhill.

A last bend is dominated by the huge Diageo distillery site (J&B, Bells, Johnnie Walker) with its chimney and pylons, then the canal comes to the impressive reach of Spiers Wharf, for many years the end of the city branch of the canal. Notable restoration work has been done on what, a century ago, were thriving sugar works, grain mills, breweries and bonded warehouses, some seven stories high. (The back walls, out of sight, are only brick!) At the far end is an elegant porticoed Georgian building of 1812, which was the original Forth & Clyde Canal Company offices. The Swifts departed from here for Falkirk (and, from 1822 with the Union opening, to Edinburgh) and later it was the city base for the pleasure craft plying out to rural Craigmarloch. Over 2005-2006 the section from Hamiltonhill to Spiers Wharf was drained in order to survey the fabric, the first inspection of the walls built over 200 years ago. They built well. When the canal was pushed on to Bowling this branch was also extended.

From Spiers Wharf the canal turned east to reach Port Dundas, a whole series of basins, a very real port in the heart of the city. The Clyde at that time was not dredged so the city was thus reached by canal. Despite being so heavily committed to commerce and industry, remarkably little was written or pictured about Port Dundas.

Even that was not the last of the story, for Port Dundas was linked to the Monkland Canal in 1793, which ran out of Glasgow to the east and was built largely to make cheap

Working on the new link to Port Dundas, November 2005.

The old bascule bridge in the newly flooded Port Dundas, 2006

Lanarkshire coal available to the city—80,000 tons of it
in 1808. The Monkland 'water' was a useful bonus for
the Glasgow Branch and the descent to the Clyde and,
even though the Monkland Canal has been filled in, the
water supply is maintained. The distillery at Port Dundas
claimed to be the largest in the world, and still dominates
the scene.

Port Dundas slowly decayed, the abandoned trenches a ready dump for rubbish, the link to Spiers Wharf culverted, the M8 slicing through to the south... But the dream was always there, and early summer 2006 sees another 'link' (a £7 million link) completed: Spiers Wharf with Port Dundas. A lock drops 4.5m (15ft) to a large kidney-shaped show basin which will be well-seen from the M8. A 70m cascade spills over to catch the motorists' eye. Going under a revamped Craighall Road, another lock leads up to connect with Port Dundas. New road lines and roundabouts encompass the site and link it with the M8.

One old bascule bridge survived in Port Dundas and a forlorn swing bridge remains, but just about everything else will have to start anew, an envisaged mix of industrial, commercial, recreational and residential facilities round the redeemed basins and the imaginative link. We may yet sit at café tables on a waterside plaza looking over the M8 to the spires and towers of the 'dear green place'.

It is interesting that while Edinburgh never stopped arguing about just where the Union Canal should end, Glasgow traders and commercial enterprises (albeit with some argy-bargy as well) were determined to have a branch right into the city centre. But could they have envisaged the site today?

The name Port Dundas commemorates one of the major backers of the canal, Sir Lawrence Dundas, a merchant who made his money selling stores to HM Forces, had estates

The Port Dundas basin in the heart of Glasgow in the winter of 1850. (© British Waterways)

and interests at Grangemouth and elsewhere, and made a killing out of the resulting developments. He cut the first sod at Grangemouth in 1768. To recap: by 1773 ships could operate to Kirkintilloch, by 1775 to Stockingfield and by 1777 the Glasgow Branch was operational as far as Hamiltonhill. However funds had run out, and it was 1784 before a government advance came through. In 1786 operations commenced to push the main line through to Bowling and everything was operational in 1790.

The canal as a human transport facility came more slowly. The world's first practical steamer, the *Charlotte Dundas*, was introduced in 1801, and in 1828 the *Cyclops* (based on a Mississippi steamboat) was tried, but both damaged the banks. The twin-hulled *Swift* came in then, but it was 1831 before design and function succeeded with the *Rapid*, the first of a whole series of 'swifts' with names like *Velocity*, *Gazelle*, *Dart*, *Gleam* and *Swallow*. A cabin, with entertainment and comforts, was a big advance on carriage travel. The *Charlotte Dundas* eventually became the canal's first steam dredger, and operated as such for many years.

In 1875, George Aitken began a goods and passenger service between Port Dundas and Castlecary (a one-legged fiddler entertained) but he was drowned in the canal a few years after. His son James launched the *Fairy Queen* 13 years later, and it, and its companions and successors, became immensely popular. An advertisement in 1916 offered a whole day's excursion: sail, dinner in the Bungalow at Craigmarloch, time ashore, 'dainty' afternoon tea on deck or in the saloon, all for four shillings and three pence (22½p)! Dinner at Craigmarloch cost two shillings (10p). The service lasted till World War Two, when the *Gypsy Queen* headed in the other direction, to Dalmuir, to be broken up.

The canal itself never recovered from the war, and in 1963 was officially closed. It rapidly suffered vandalism and dumping and, with fatalities, voices were raised to fill it in. Happily, at the last moment, its worth as a leisure asset became better understood and, ever since, slowly and determinedly, it is being brought back to life. We enjoy it as a pleasant walk only because of the efforts of many enthusiasts over the years.

16
DOWN TO THE CLYDE:
MARYHILL TO BOWLING

OSLR 64; OSE 342

From the Stockingfield Junction a five minute walk leads to Lock 21, the western end of the summit pound that we have followed ever since Wyndford Lock, what feels a long time ago. Maryhill Road runs under the canal at this point, and there is staging and a slip before Lock 21, which is pretty well unused. The state of the old canalside inn, an original hostelry, might indicate why. There is still the Kelvin Dock pub across Maryhill Road, however, and Lock 27 (the pub) is not far ahead. The name Maryhill comes from an impecunious heiress, Mary Hill, who freed land for development with the condition her name was commemorated. Chemical works, timber yards, metal fabrications and ship building developed.

The Maryhill Locks form perhaps the most spectacular flight on the Lowland canals as it drops from the summit level, Lock 21, to Lock 25 and the Kelvin Aqueduct just beyond, with big ink-blot shaped basins between each drop and historic Kelvin Dock lying off at a tangent from Lock 22. Kelvin Dock was operational from 1789 to 1949, the oldest yard on the canal, where canal company boats were launched, both sideways and stern first; and there was also a dry dock.

The smaller barges were called 'scows' (perhaps from the Dutch *schouw*, a flat-bottom boat) and the large barges were 'lighters'. Most notably, here the first-ever 'puffer' was built

Opposite: The Maryhill flight of Locks (© Dr Patricia Macdonald)

(the *Glasgow*), a type of coastal cargo vessel immortalised in the Para Handy stories of Neil Munro. The puffer evolved from the scow, which was originally towed by horses, till engines came along. Landing craft for the D-Day landings in World War Two were one of the last projects undertaken in the docks. Naval ships have not been great canal users, though a miniature submarine in 1952 caused some interest.

The four-arched Kelvin Aqueduct, built in 1790, and then the largest project of such a nature in Britain, is still impressive. Seaton having retired, Robert Whitworth was the engineer called to take the canal the 400 feet (130m) across the 70 foot (22m) deep River Kelvin valley. The technology of the times necessitated the aqueduct's massive strength whereas the 'great three' of the Union Canal (Avon, Slateford, Almond) were able to use newer skills to produce more slender structures. The cost was £8,500, against an estimate of £6,200, so over-running is nothing new, though more justified with such a pioneering feat which brought tourists by their thousand and inspired plenty of well-forgotten poetic odes! One history of Glasgow enthuses, ' ...uniting the German and Atlantic Oceans...

Maryhill locks and the Kelvin Aqueduct leading into the trees

square rigged vessels are sometimes seen navigating 70 feet above spectators… a pre-eminence over everything of a similar nature in the Kingdom'.

The Kelvin Walkway runs below the aqueduct, linking Kelvingrove with the West Highland Way. At the east end of the aqueduct a path (with steps) drops down to gain the walk/cycle way. The great strength of the aqueduct is best seen from below despite the choking of trees (winter is a good time to see it) and a return should be made by the same route to continue along the canal. The Glasgow Botanic Garden, University, Kelvingrove Museum and Art Gallery, Park and Hunterian Museum can all be reached downstream by the walkway. This whole area is scheduled for renewal, but this is not likely in the lifetime of this guide.

Considering the canal is in the middle of a city, the scenery and setting remains remarkably rural. The first bridge, Govan Cottage Bridge, was rebuilt as part of the restoration in the Millennium style. Up on the south, the cottage was built in railway style at the time the railways owned the waterway. Kelvindale Station lies above the bridge. Walking on, the view is dominated by two gasometers which the towpath passes. There are signs of old mineral line crossings, and Lock 26 has current railway tunnels passing underneath just to east and west. Lock 27 is interesting. An original bascule bridge carried the Crow Road (North) over the canal here at Temple, but in 1932 the Bearsden Road was realigned and a huge steel lifting bridge installed. This has since gone and there's now a sturdy four-lane, iron girder bridge. A model of the lifting bridge is in the Museum of Scotland in Edinburgh (Chambers Street). A footbridge crosses the canal just west of Lock 27.

Lock 27 is also a modern public house, on the site of an original lock-keeper's cottage, with plenty of outside seating beside the canal. You can take your pick for the distance to Bowling; one sign says 7¾ miles. Another 8½. The nearby Canal Restaurant has closed. Here and on the

Recognisable still—Lock 27 with timber maturing.

stretch ahead new housing has been developed on the site of the once busy timber yards, timber brought from across the North Sea and through Grangemouth.

Continuing, there is a massive iron bridge with red sandstone abutments, the busy Bearsden Road. Beyond there is staging and facilities for canal boats, kept locked out of necessity.

Attractive houses face the canal and a utilitarian concrete bridge serves the Lynch Estate. The delightful Netherton swing bridge lies beside housing akin to the south of England perhaps. Last time I was here a cormorant went winging past. Gentle suburbia continues with houses ranged above the railway on the right. The Westerton footbridge, the next landmark, a single girder bridge with a twirl down at the south end, leads to Westerton station, and stations from here to Bowling can make useful returns for the parts of canal walked.

Another attractive flight of locks, 28-32, Clobberhill, leads to a footbridge with the Blairdardie playing fields to the right and a small tree area on our side pleasant to walk through. There's a big new double concrete road bridge for the Great Western Road. The historic Bard Avenue bascule bridge has a wee shop close by. (Bascule comes from the French word for a see-saw.)

Clobberhill Locks

The Boghouse Locks, 33-36, give a last flight, with only three more locks to come before meeting the sea. Much had to be rebuilt in this area to reopen the canal, even parts of the waterway that had been retained had been filled in to a couple of feet deep 'for safety'. Lock 36 had to be re-instated, hence the odd kink in the approach to it. Duntreath Avenue Bridge is another big concrete structure connecting Yoker and Drumchapel, where we begin to be aware of the Kilpatrick Hills. The bridge was culverted before the millennium restoration. In Victorian times this was pleasant countryside, Drumchapel a holiday destination. The Linvale bascule bridge and the long straight to the Clyde Shopping Centre suffers badly from misuse, the water a dump for plastic bags, bottles, carry-oot cartons and shopping trolleys—a section to hurry through, though there is the temptation of refreshments! After the Argyll Road bridge there are ranks of superstores, shopping malls and the Lowry-busy Sylvania Way footbridges, quite an alien shock. With Glasgow Airport across the river, planes fly over constantly. A Victorian bandstand cringes in a corner. The footbridges over the canal can lift one at a time so as not to impede the flow of shoppers!

Clydebank was devastated by wartime bombing, when 4000 homes were destroyed, with heavy civilian casualties, and the steady decline in shipbuilding has done little for morale. Gone are the days when ships like the *Lusitania*, *Queen Mary* and *QE2* towered beside the Clyde. Now, fixed in concrete, is the *Debra Rose*, McMonagles Sail Thru Fish and Chips Take Away. Starlings have found a niche among the human ebb and flow.

The Kilbowie Road Bridge, as we leave, had to be rebuilt as it was previously culverted. Looking back, the tower of the Clydebank Municipal Buildings can be seen. The storms of 1968 saw the angel on top, which came from the 1901 Glasgow exhibition, removed. The north side of the canal has a high concrete wall running along to an abutment of an old bridge. A railway runs alongside on the left, with a branch bearing off into the docks, and this soon passes below the canal, the tunnel only noted from the towpath as there's a brick wall. The view along the line leads the eye back to the cranes of Clydebank. The huge Singer Sewing Machine Factory lay north of here, now a modern industrial estate. In its heyday the cooling water flowing into the canal supported a thriving number of goldfish. At the height of its success over a million sewing machines were produced in a year. The works closed in 1980.

Shortly after, the canal crosses a minor road. Steps descend onto Boquharan Road. It is worth going down to see two things: the unusual narrow brick bridge with its raised pedestrian way and, across the Dumbarton Road, on the wall above the Park Tavern, an interesting carved feature showing a First World War battleship.

Trafalgar Street pedestrian bridge (all metal mesh) comes next, an area favoured by swans. One October I spent an enjoyable time watching them from the bridge as they indulged in vigorous preening, often rolling over upside down in the water and flapping violently, the cygnets testing out their new-discovered powers of flight.

Dalmuir Drop Lock.

Asters still gave a touch of blue on the green banks, while the trees of this very woody stretch were lighting up with autumn colours.

One of the most ingenious tricks of the Millennium Link comes next: the Dalmuir Drop Lock. Originally a bascule bridge sufficed on the Dumbarton Road, but later there was a sturdy swing bridge across which tramcars clanked. Both trams and canal ceased to function in the sixties; Glasgow's very last tram was the No 9 Auchenshuggle to Dalmuir West. Such a major road across the canal line set problems, the solution being to create a lock that can lower boats under the obstruction and then raise them again—the first such lock of this kind in Britain; nothing distinctly different to look at, but brilliant.

Continuing, there's an almost graffiti-free bascule bridge (Farm Road) with huge high pylons crossing the view, for they span the River Clyde. From the large 1934 Erskine swing bridge a road leads down to the slip where the ferry operated across the Clyde before the 1971 bridge opened. Do go down, there's a great view below the arc of bridge to the river and the wooded crags towards Dumbarton.

In the 19th century a ferryboat could carry up to 40 head of cattle and was pulled across on a chain. Now, as one guide puts it, the bridge is 'elegant in the distant landscape but awesomely monstrous above the village'. Ironically, this last section (almost as level as the Union Canal) was soon made unnecessary by the Clyde being regularly dredged to allow sea-going ships up into the city. Had practical dredging come earlier, the Forth & Clyde canal might have ended in the city below Port Dundas.

Returning, look out left, for the entrance to The Saltings, a nature reserve, where there's a notice board and car park, the site dominated by the great bridge. One can exit from the reserve back to the towpath 200 yards west of the swing bridge. From the swing bridge, about the same distance along the main road is the village cluster of shops, post office, coffee bar and the Glen Lusset Restaurant. The restored Lock 37 lies just west of the swing bridge.

Old Kilpatrick's church tower is a landward mark. The wartime blitz has left some of the gravestones pitted, chipped or repaired from enemy machine gun fire. Ferrydyke bascule bridge is the only break in a mile long stretch of canal. A bridge keeper's cottage has survived. Nearby was the western terminal fort of the Roman Antonine Wall, which descended to the Clyde here after its sweep across Scotland from the Forth. There's an aptness in the two great engineering feats, almost a millennium apart, marching across the country so companionably. The Ferrydyke name could conceivably have originated then.

On a winter walk this way I once watched two young swans fly in to land on the canal; they presumed it was water and it was only when they put down their big feet they discovered the canal was frozen solid. The sight of swans sliding along on their bottoms, on their necks, and round and round reduced spectators to hysterics. The attractive Kilpatrick Hills are close to the canal now and a railing shows where a burn, coming off the slopes, passes under the canal. Suddenly you come on the final features of the canal.

Erskine Bridge from the old ferry slip.

Lock 38 leads into a big high-walled basin below the white house, with a last bascule bridge at its end before going under the once-operational Caledonian Railway swing bridge (Lock 39) to the boat-crowded last basin itself. This can be busy and, in winter, many boats are moored there for the season with the yachties busy working on them at weekends, always a very attractive scene. It took 22 years for the canal to reach Bowling (in 1790). I trust we've had a less fraught experience on our journey across Scotland.

Circumambulating the basin you come onto an over-flow, which was once the original sea lock. Between two and five million gallons of water flows through the Forth & Clyde canal system each day. The view up-river is always fine, with the Erskine Bridge seen at its best. Cross the lock which leads out to Bowling Harbour. On the point down-river is a monument to Henry Bell, of the *Comet* steamship fame. (The *Comet* was wrecked on Craignish Point in 1820 while operating a Glasgow-Fort William service through the Crinan Canal.) Continuing, the trim 18th century Customs House is passed before you go under the abandoned railway again to arrive back at the bascule bridge.

Your canal walk is over. To celebrate the opening of the full canal in 1790, a hogshead of Forth water was poured into the Clyde. Your celebrations will no doubt be more modest.

And the last practicalities: to leave, take the track from the bascule bridge left (not the track tight against the railway arches, that's the vehicle access to the harbour basin) to climb up and over the busy railway line to reach the Bowling road (A814). Across the road there's a bus stop for buses back to Clydebank/Glasgow, a service operating about every quarter of an hour. For trains, a five minute walk west along the main road leads to the station. A car park is passed on the way, overlooking the Old Bowling Basin, and visiting cars should be parked here and not down by the canal basin.

The Companion again:

'Before concluding, let it be observed, that there are few foreign landscapes whose beauties really excel those of the scenes just described. True, some countries, enlivened by the rays of a more vertical sun, can boast of richer fruits, and more luxuriant verdure, mountains more lofty even than those of Caledonia, but there are few landscapes that unite the cultivated charms, the rural scenery, and the lofty mountain grandeur, which are so finely blended as those of the Forth & the Clyde.'

Opposite: Looking along the Falkirk Wheel aqueduct

APPENDIX:
PRACTICAL INFORMATION

Contact numbers with relevance to the Union Canal and Forth & Clyde Canal. As facilities change constantly or keep variable opening schedules, a telephone call in advance is always recommended.

Accommodation can be booked through the Tourist Information Centres (TIC) at Edinburgh, Linlithgow, Falkirk, Glasgow, Dumbarton, as listed, or via the National Call Centre: 01506-844-600.

Action Outdoors	0141-354-7540
Adventure Centre, Ratho	0131-333-6333
Almondell & Calderwood Country Park	01506-882-254
Almond Valley Heritage Centre (shale museum) Livingston	01506-414-957
Alvechurch Waterways Holidays (Falkirk Wheel)	0121-445-2909
Antonine Walkway Trust	01236-827-242
Beecraigs Country Park	01506-844-516
Beefeater Rosebank (Camelon)	01324-611-842
Black Prince Holidays (Falkirk Wheel)	01527-575-115
Boating Holidays (Falkirk Wheel)	01324-880-280
Boathouse Restaurant (Falkirk)	01324-613-311
Bonny Barge (Lock 16)	07720-866-397
Bridge 19-40 Canal Society (Winchburgh)	01506-417-685
Bridge Inn, Ratho	0131-333-1320/1251
British Waterways (Lowland Canals), New Port Downie (Falkirk Wheel)	01324-671-217

British Waterways Scotland,
  Canal House, 1 Applecross Street,
  Glasgow, G4 9SP   0141-332-6936

Cairnpapple Hill (Historic Scotland) 01506-634-622
Callendar House   01324-503-770
Canal Inn (Camelon)   01324-612-597
Capercaillie Cruisers (Falkirk Wheel) 0131-449-3288
Carron Sea Lock   01324-483-034
Colzium House   via 0141-304-1800
Craft Daft (Studio Boat,
    Glasgow Bridge)   07958-198-034
    or   01292-280-844
Cross Scotland Holiday Cruises   0131-333-1424
Cygnus (narrowboats, hiring, cruises,
    cycle hire: Auchinstarry based)   0131-445-3388

Debra Rose (Clydebank)   0141-951-1333
Dumbarton TIC   01389-763-444
Dumbreck Marsh Nature Reserve   01236-780-636

Edinburgh Canal Centre (Ratho)   0131-333-1320
Edinburgh Canal Society
    (City Cruising/Charters)   0131-556-4503
Edinburgh Castle   0131-225-9846
Edinburgh TIC   0131-473-3800
Ettrick, The (below Erskine Bridge) 01389-872-821

Falkirk & District Canal Society   01234-622-997
Falkirk TIC (all year)   08707-200-614
Falkirk Wheel   01324-619-888
Falkirk Wheel (Trip Boats)   01324-619-888
Forth Canoe Club (Harrison Park) 0131-229-9586
Forth & Clyde Canal Society
    (Glasgow Bridge)   0141-772-1620

Geo Projects (Canal Maps)   0118-939-3567
Glasgow TIC   0141-204-4400
Glen Lusset Restaurant
   (below Erskine Bridge)   01389-872-727
Govan Seagull (Falkirk, Bantaskine)   01324-621-733
Grangemouth Museum   01324-483-291

Kirkintilloch; Auld Kirk Museum   0141-578-0144
Kirkintilloch; William Patrick Library 0141-776-7484
Kirky Puffer, (Wetherspoon)
Townhead, Kirkintilloch   0141-775-4140

Leisure Dome, Bishopbriggs   0141-772-6391
Linlithgow Palace   01506-842-896
Linlithgow Story (museum)   01506-670-677
Linlithgow TIC (Apr-Sept)   01506-844-600
Littlemill Inn, Bowling   01389-875-833
Lock 27 (Restaurant, Glasgow)   0141-958-0853
LUCS (Linlithgow Union
   Canal Society) operates *Victoria*,
   St Magdalene, etc   01506-843-194
   Tearoom (LUCS)   01506-671-215

Mackay Seagull (Ratho)   0141-333-1320
Magic Cycles (hire; Bowling)   01389-873-433
Marine Cruises (Falkirk Wheel)   01691-774-558
Maryhill (replica puffer, cruising)   0131-449-3288
McMonagles Fish Restaurant,
   Debra Rose, Clydebank   0141-951-2444
Muiravonside Country Park   01506-845-311
Museum of Scotland, Edinburgh   0131-247-4422

National Gallery (Edinburgh)   0131-642-6200

Ogilvie Terrace Moorings   01324-671-217

Park Bistro   01506-846-666

Peccadillo (barge—Hamiltonhill)	0141-332-1242
Pride of Belhaven (Ratho)	0131-333-1320
Pride of the Union (Ratho)	0131-333-1320
Railway Inn, Bowling	01389-800-234
Ratho Adventure Centre	0131-229-3919
Ratho Princess (charter; Ratho)	0131-333-1424
Royal Botanic Gardens, Edinburgh	0131-552-7171
Seagull Trust (Disabled cruises)	0131-229-1789
Scottish Inland Waterways	
Association	01786-870-501
St John Crusader II (Ratho)	0131-333-1320
St Magdalene (see LUCS)	
St Michael's Church (Linlithgow)	01506-842-188
Stables, The (Restaurant)	0141-777-6088
Standing Waves Leisure (Canoeing;	
Bonnybridge)	01324-810-888
Star & Garter Hotel, Linlithgow	01506-843-362
Tally Ho Inn, Winchburgh	01506-890-221
Thistle Down (Narrow boat hire;	
Linlithgow)	0131-621-0950
Underwood (Lock 17) Restaurant	01324-849-111
Union Inn (Lock 16)	01324-613-839
Water of Leith Visitor Centre	0131-455-7367
Waterways Trust Scotland	01324-677-820
William Patrick Library, Kirkintilloch	0141-776-7484
Victoria (see LUCS)	
Yarrow Seagull (Kirkintilloch)	0141-777-7165
Zazou (Restaurant, cruising, Edinburgh)	0131-669-5516

CANAL COUNTRY CODE

Good manners and friendliness are the basics to ensure safety on canal towpaths. Think of other users. The Country Code is a charter of freedom, not a restriction.

Canal bye-laws do not allow horse-riding, motorbike or vehicle use of the towpath. This is obviously in the interests of safety as well as enjoyment.

Cyclists should take care not to startle walkers (coming from behind or round blind corners) and treat aqueducts and locks with great care, as indeed should pedestrians, who should also let fishermen know of their approach. Anglers should be careful not to interfere with other users. Beware of overhead power-lines.

Close all gates. There is nothing calculated to annoy farmers more than having to round up strayed livestock. Don't go over walls, or through fences or hedges. There are few places without gates or stiles where needed on our route.

Leave livestock, crops, boats and machinery alone.

Guard against all risk of fires.

Dump your litter in bins, not in the countryside or the canal. You'll see some bad sights, don't add to them. Poly bags can mean death to a grazing cow, broken glass is wicked for both man and beast, and drink cans an insufferable eyesore.

Leave wildlife alone. Collect memories, not specimens.

Walk quietly in the countryside. Nature goes unobtrusively and you should too.

Dogs need strict control and should not be allowed to foul the towpath.

Be extra careful when walking on roads, however quiet these appear. Manic drivers are no respecters of pedestrians.

Make local contacts. Rural people are still sociable and a 'crack' will often be welcomed. Use the Tourist Information Centres and bookshops along the way to widen knowledge and enrich your experience. Local people met in bars and cafés are often a fund of information.

ACKNOWLEDGEMENTS

The many people who contributed to the first edition are again warmly thanked, as are others more recently involved, many questioned on site who remain anonymous. A special debt to British Waterways Scotland is acknowledged, many of whose staff answered queries and, in the case of the Chief Engineer George Ballinger, brought the recent link from Spiers Wharf to Port Dundas to life for me. Other staff, past and present, I must mention are Jill Richards and Helen Rowbotham. Dr Patricia Macdonald, expert aerial photographer, is thanked for the superb shots of the Kelvin Locks. Other enthusiastic experts who helped are Guthrie Hutton, Anne Street, Ronnie Rusack, Mel and Judy Gray (of the Linlithgow Union Canal Society), Bill and Sandra Purves (Edinburgh Canal Society) and Captain Gordon Daly (Seagull Trust). Thanks also to Don Martin and others at the William Patrick Library, Kirkintilloch, and the Edinburgh City Library and to Jill Adam, who walked some wintry days, Sheila Gallimore, who typed my notes so effectively and those at Mercat Press who took on the practicalities of publishing the guide.

All facts have been checked as far as possible, but the author and publishers cannot be held responsible for any errors, however caused or for the use to which the guide is put. Developments and changes occur continuously, and the publishers would welcome information for correcting and updating any future edition.

BIBLIOGRAPHY

Local libraries always order books for readers if requested. This is a selected list, but each book itself will lead on to others. The Tourist Information Centres on or near the route are a useful source of new local booklets.

Anderson, R: *A History of Kilsyth,* Duncan, 1901.

Anton, P: *Kilsyth, A Parish History,* John Smith, 1843 (reprinted).

Bowman, A I: *Kirkintilloch Shipbuilding,* Strathkelvin District Publications, 1983.

Bowman, A I: *Swifts and Queens: Passenger Transport on the Forth & Clyde Canal,* Strathkelvin Dist Pubs, 1984.

Bowman, A I: *The Gipsy o'Kirky,* Strathkelvin Dist Pubs, 1987.

Buildings of Scotland series (Penguin) has comprehensive coverage of Edinburgh, Lothian and Glasgow.

British Waterways Scotland: *Exploring Scotland's Lowland Canals and the Millennium Link Project,* 2003.

British Waterways Scotland: *The Forth & Clyde Canal, the Union Canal and the Falkirk Wheel.*

Companion for Canal Passengers Betwixt Edinburgh and Glasgow, 1923. Facsimile, LUCS booklet, 1981.

Dowds, T J: *The Forth and Clyde Canal. A History,* Tuckwell, 2003.

Forth & Clyde Canal Society: *The Forth & Clyde Can Guidebook,* East Dunbarton Libs, 2001 (periodically revised since 1985).

Haldane, A R B: *The Drove Roads of Scotland,* David & Charles, 1973 *et seq*. (a classic).

Hanson, W H and Maxwell, G S: *Rome's North-West Frontier— The Antonine Wall,* EUP, 1983.

Hendrie, W F: *Discovering West Lothian,* John Donald, 1986.

Hendrie, W F: *Linlithgow, Six Hundred Years a Royal Burgh,* John Donald, 1989.

Horne, J. (edit): *Kirkintilloch,* 1910 (reprint 1993).

Hume, J R: *The Industrial Archaeology of Scotland. 1: The Lowlands and Borders.* Batsford, 1976.

Hutchison, J; Weavers, Mines and the Open Book, Cumbernauld, 1986 (Kilsyth history).

Hutton, G: *Scotland's Millennium Canals,* Stenlake, 2002 (Comprehensive history).

   A Forth and Clyde Canalbum, Stenlake (old pictures).

   Forth and Clyde, The Comeback Canal, Stenlake (old pictures)

Johnstone, Anne: *The Wild Frontier,* Mowbray House, 1986
    (large format illustrated introduction to the Roman Wall).

Lawson, L: *A History of Falkirk,* Falkirk, 1975.

Lindsay, J: *The Canals of Scotland,* David & Charles, 1968.

Love, D: *Scottish Kirkyards,* Hale, 1989.

Martin, D: *The Story of Kirkintilloch,* Strathkelvin Dist Pubs, 1980.
    The Forth and Clyde Canal: A Kirkintilloch View,
    Strathkelvin Dist Pubs, 1985.

Martin, D & Maclean, A: *Edinburgh & Glasgow Railway*
    Guidebook, Strathkelvin Libraries, 1992.

Massey, A: *The Edinburgh & Glasgow Union Canal,* Falkirk
    Museum, 1983.

Robertson, A J: *The Antonine Wall,* Glasgow Archaeological
    Soc, 1990 (useful field guide).

Scott, J: *The Life and Times of Falkirk,* John Donald, 1994.

Skinner, B C: *The Union Canal, A Report,* LUCS, 1977.

Smith, R: *25 Walks, Edinburgh and Lothian,* Mercat Press, 2006.

(Old) Statistical Account of Scotland: Edited by John
    Sinclair 1791-99 and reprinted in 1980s by E P. Publishing
    by areas; Volume 2 covers the *Lothians*; Volume 9 covers
    Dunbartonshire, Stirlingshire and Clackmannanshire.

Thomas, D St J: *Forgotten Railways, Scotland,* David & Charles.

Tranter, N: *Portrait of the Lothians,* Hale, 1979.

Waldie, G: *A History of the Town and Palace of Linlithgow,*
    1897/1982.

Watson, T: *Kirkintilloch: Town and Parish,* John Smith,
    Glasgow, 1894.

Williamson, A G: *Twixt Forth and Clyde,* London, 1944.

Willsher, B: *Understanding Scottish Graveyards,* NMS, 2005.

INDEX